The Nail File

Related Macmillan titles

Manicure, Pedicure and Advanced Nail Techniques by Elaine Almond

Science and the Beauty Business by John Simmons
Volume 1: The Science of Cosmetics
Volume 2: The Beauty Salon and its Equipment

Hairdressing – The Foundations by Leo Palladino

The Principles and Practice of Hairdressing by Leo Palladino

Mastering Hairdressing by Leo Palladino

The Nail File

LEO PALLADINO
JUNE HUNT

First published 1992 by
MACMILLAN EDUCATION LTD
Houndmills, Basingstoke, Hampshire RG21 2XS
and London
Companies and representatives
throughout the world

Filmset by Wearside Tradespools, Boldon, Tyne and Wear

Printed in Hong Kong

British Library Cataloguing in Publication Data
Palladino, Leo
The nail file.
1. Man. Nails
I. Title II. Hunt, June
612.799
ISBN 0-333-52584-1

10 9 8 7 6 5 4 3 2 1
01 00 99 98 97 96 95 94 93 92

Contents

Acknowledgements

The authors would like to thank Rosemarie Tait, Janet Sims and Abraham Moss for providing most of the illustrations, and for Rosemarie's invaluable support and encouragement in preparing the book.

The chapter 'Guidelines to Hygienic Manicuring' is an adaptation of *Guidelines to Hygienic Hairdressing*, Professor N. D. Noah, Department of Public Health and Epidemiology, King's College School of Medicine and Dentistry, and is published with kind permission.

Part One

Nail File

UNIT 1

Introduction

Attractive, well cared for hands and nails are an essential feature of personal presentation for both men and women.

Throughout history, manicure (*manus*: hand, *cura*: care) has been practised by many civilisations – the Egyptians, Greeks, Romans – right up to the present day. In the past it was only the ruling classes who enjoyed the luxury of manicure. Well cared for hands clearly showed a person's status, revealing that they never did any manual work.

Today, of course, manicure is widely practised in all sections of society and has developed into a substantial industry.

This book deals not just with established manicure techniques but also includes the latest advances in nail technology.

For convenience, the book is divided into two parts. Nail File deals with the Art of Manicure and gives all the practical techniques and procedures. Fact File covers the Science of Manicure and gives the background theory.

Unit 2

Preparing for Manicure Treatments

THE MANICURE ROOM

The room or area used for manicure treatments must meet the following requirements:

1 Hot and cold water must be readily available. This will be required at various stages as follows:
   - For washing hands before and after dealing with the client
   - For soaking the client's nails to soften cuticles, making them easier to push them back and helping the massage cream to penetrate
   - For removing dirt and debris from the free edge of the nail
   - For altering the bevel of artificial nails
   - For removing oils and caustic preparations – cuticle remover – from the client's nails
   - For cleaning surfaces, trays, etc.

2 The furnishings and surroundings must meet professional standards. These include:
   - A spotlessly clean area is the ideal
   - Floors should be covered in easily cleaned materials. There should be no cracks or tears as these could be dangerous
   - Decorations should be uncluttered and hygienic
   - Work tables must be the correct height for comfortable working and there must be no dirt traps to allow bacterial growth. It is essential that work tables are easy to clean

- An autoclave should be available to sterilize all metal tools. Some plastic materials can be damaged by the temperatures of high-pressure steam so check all instructions. Sterilization prevents cross-infection with bacteria, viruses and fungal parasites
- All waste, including soiled linen, must be placed in covered bins
- Storage of materials must be safe
  * A metal cabinet must be available for flammable materials such as solvents like acetone and alcohol. Flammable materials should only be stored in small quantities because of the danger of fire
  * All products should be stored upright to prevent spillage and kept in a cool, dark, place
  * A fire extinguisher should be available

Layout of manicure table

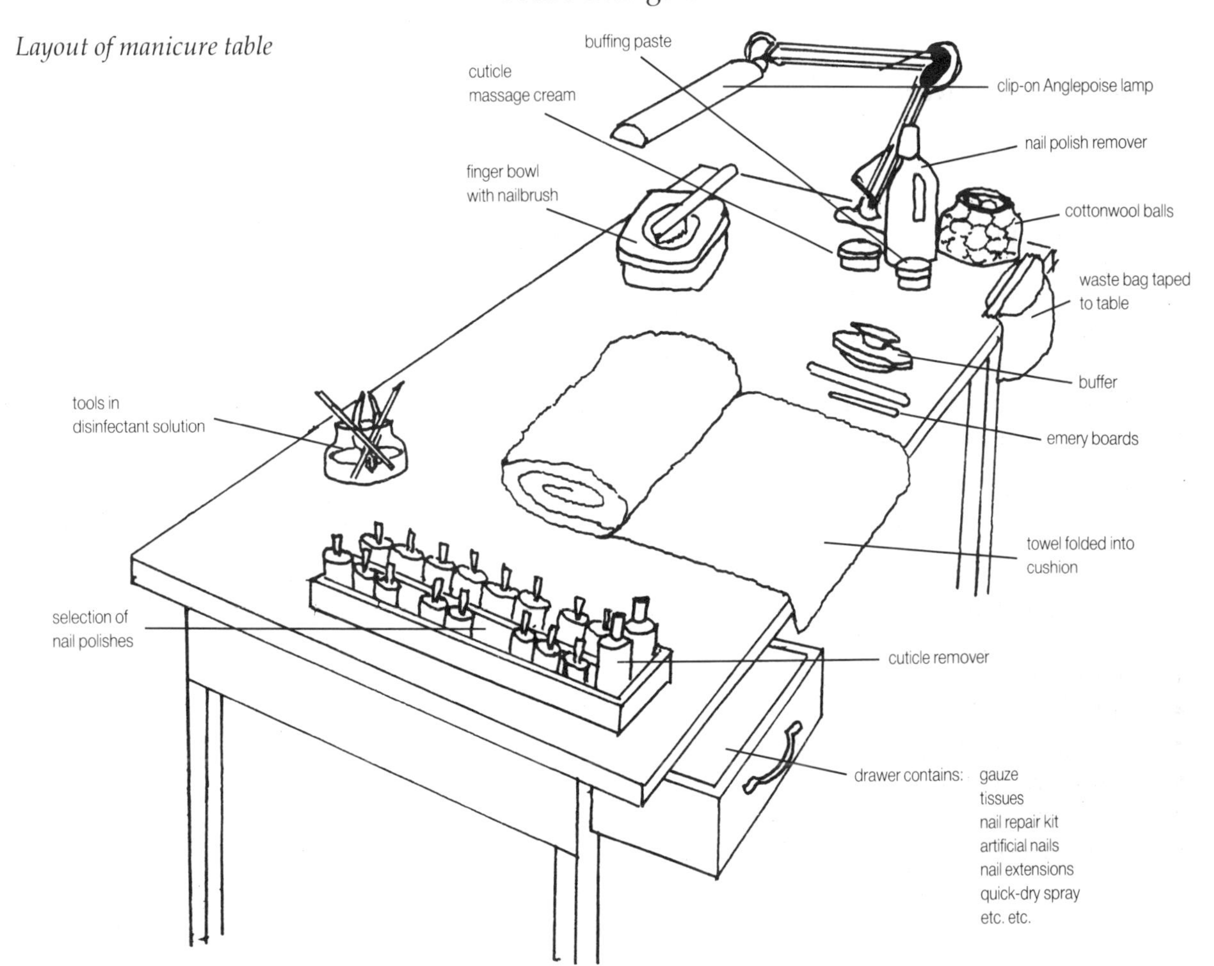

*COSHH regulations – Control of Substances Hazardous to Health.

- Adequate lighting must be available for detailed work
- Ventilation should be able to deal with the fumes of the various materials. Some of these can be dangerous if inhaled.*

3 All preparations for the manicure should be completed before the client arrives. This avoids keeping the client waiting unnecessarily. The manicurist should carry out the following:
   - All tools and materials must be assembled in position on the manicure table for easy access
   - A wide selection of cream and pearl polishes should be available. These should be checked for correct consistency (see p. 32)
   - Sterile tools (see checklist below) should be laid out on white tissue.

Checklist for Manicure

Tools

- Cuticle nippers
- Cuticle knife
- Nail scissors
- Rubber-ended hoof stick
- Orange sticks
- Buffers – one leather, one four-sided
- Emery boards and discs
- Spatula

Equipment

- Water bowl designed for soaking fingers
- Waste bin, bag or bowl
- Double-sided adhesive tape
- Autoclave
- Covered towel and waste bin
- Manicure cushion
- Table and chairs at the correct height
- Cottonwool in containers

*COSHH regulations – Control of Substances Hazardous to Health.

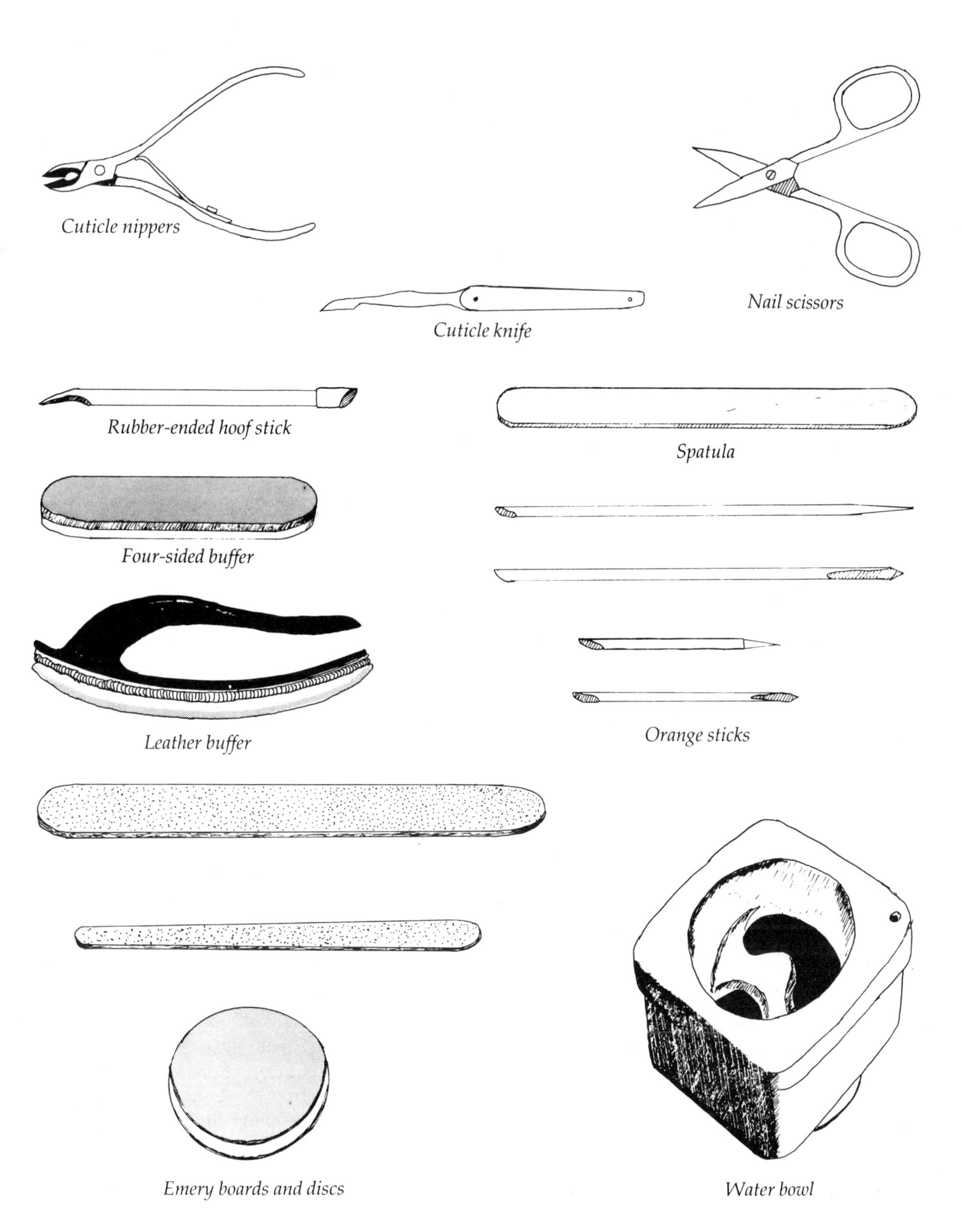
Cuticle nippers
Nail scissors
Cuticle knife
Rubber-ended hoof stick
Spatula
Four-sided buffer
Leather buffer
Orange sticks
Emery boards and discs
Water bowl

Materials

- Absorbent surgical cottonwool
- Nail-repair tissues, silk and linen fabric
- Nail-mending glue
- Selection of false nails and associated items
- Large colour selection of cream and pearl polishes
- Base coat and top coat
- Quick-drying spray
- Nail strengtheners
- Buffing paste
- Cuticle massage cream and cuticle oil
- Hand cream
- Cuticle remover
- Tissues

Manicure Cosmetics

The following table lists the manicure cosmetics likely to be used. It names the product, its main ingredients, its uses and effects.

Product	*Main ingredients*	*Uses and effects*
Polish remover	amyl/butyl/ethyl acetate, acetone, glycerol	Solvents for polish/enamel/ varnish. Contains oil because solvents dry the nail plate
Buffing paste	'jeweller's paste' – stannic acid, kaolin or talc	Mild abrasive, shines nails. Gives smooth base for polish application. Action improves circulation to matrix
Cuticle massage cream	Oil in water emulsion, glycerol, white iodine, beeswax, mineral oils	Emollient, softens and nourishes cuticles. Used in preparation for further treatment
Cuticle remover	Sodium/potassium hydroxide	Caustic alkali, dissolves dead cuticle, softens keratin. Drying effect if over-used. Has a mild bleaching effect
Hand cream	Oil in water emulsions, lanolin, glycerol	Softens and nourishes skin. Allows easier hand massage

Product	*Main ingredients*	*Uses and effects*
Nail enamel: base coat, top coat	Nitro cellulose Formaldehyde resins – aryl sulphonamide formaldehyde Plasticizers – castor oil Ethyl/butyl/amyl acetate Pigments – usually synthetic Base coats and top coats are chemically similar though top coats are thicker	Plastic film Gloss Flexibility Solvent Colour
Pearlised enamel	Guanine from fish scales	Pearlised effects of enamels/ polishes
Nail hardeners or strengtheners	Alum, zirconium chloride, glycerol, formaldehyde resins, acrylic plastics	Hardens/thickens poor or damaged nails. Allergic reactions sometimes caused
Nail white pencils	Titanium dioxide, white soft paraffin	Used under free edge of nail to whiten
Artificial/sculptured nails	Plastic polymers and monomers, with hardeners	Nail extensions, formed on nail. Can be allergenic
Instant nail glue	Cyanoacrylates	Attaching false nails (dangers see p. 56)
Nail polish thinner	Contains no additives such as oil which would disturb balance of the product	Thinning polish
Nail primer	Methacrylate	Used for etching natural nails in preparation for artificial nails

UNIT 3

Patterns of Work

It is important to state that there is no 'correct' manicure procedure. The order and method of working depend on the needs and comfort of the client and manicurist.

But it is important to establish a normal working routine. This should be varied only in special circumstances. A left-handed client will need more time spent treating the cuticles of the left hand. This is because the left hand is the 'working' hand and will probably be in a worse condition than the right.

A left-handed manicurist will invariably find it easier to work from left to right to avoid smudging the enamel.

The routine adopted should enable the manicure to be efficiently completed in the shortest time possible.

Other changes in the routine may be as follows:

- The client may not want any enamel applied. In this case buffing should be completed at the end of the manicure.
- The client's cuticles may be in exceptionally poor condition, perhaps due to nail biting (onychophagy) in which case the skin surrounding the nail and cuticle may also be chewed. Sometimes there is excessive overgrowth of the cuticle. (Pterygium is a condition where the excessive cuticle is firmly attached to the nail plate.) Both hands will need a longer softening period. Alternatively an oil manicure may be given. This will soften the cuticle so that the manicure procedures can be carried out without discomfort. This includes pushing back or loosening the cuticles with the cuticle knife before removing the excess with cuticle nippers.

Before starting work make sure of the following.

- Jewellery is removed and the client's hands are wiped with disinfectant
- Hands and nails are inspected thoroughly, using a magnifying inspection lamp if necessary
- Tools are placed in antiseptic solution when not in use to inhibit the growth of bacteria.

SEQUENCE OF TECHNIQUES

The following procedure chart lists the most commonly used manicure techniques in logical sequence. This sequence should:

1. Make the manicure comfortable and efficient for client and manicurist
2. Allow the manicure to be completed in the shortest time possible
3. Allow the manicurist – if right handed – to work from left to right. This avoids smudging and allows completed work to be easily seen throughout. This is particularly useful for comparing the shapes of the nails and altering them by filing

This chart should be kept available while working until a routine is firmly established.

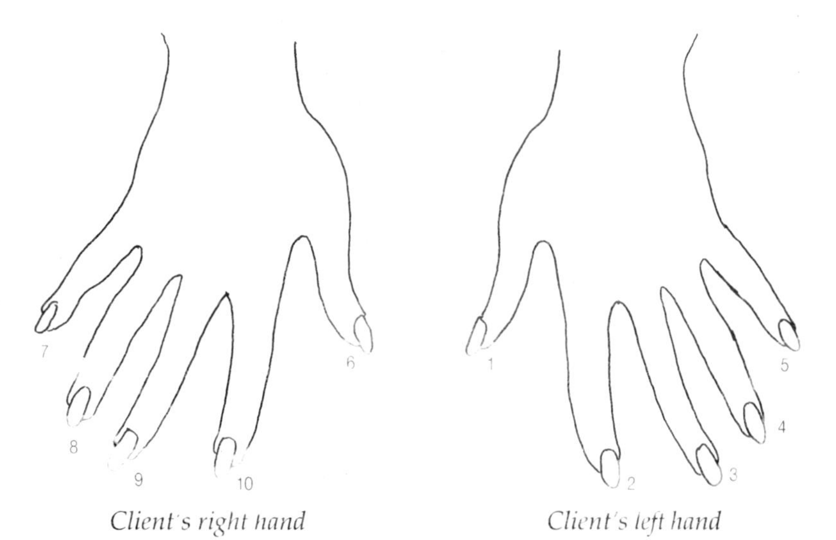

Client's right hand *Client's left hand*

Step 1

Clean *both* hands of the client with an antiseptic solution

Step 2

Inspect *both* hands of the client. Check for contra-indications to manicure (see Unit 15). Assess the condition of the nails and cuticles. Discuss any problems with the client and recommend particular treatments and products

Step 3

Remove old polish from *both* hands beginning with the left

Step 4

File left hand working from thumb across to little finger

Step 5

Buff left hand working from thumb across to little finger

Step 6

Apply cuticle massage cream to the left hand and fingers 1–5

Step 7

Place left hand in manicure bowl to soak

Step 8

File right hand working first on the thumb then on the little finger and across the hand

Step 9

Buff right hand working from 6–10

Step 10

Apply cuticle massage cream to right-hand fingers 6–10

Step 11

Place right hand in manicure bowl to soak

Step 12

Remove left hand from bowl and pat dry

Step 13

Apply cuticle remover to left-hand cuticles – fingers 1–5 – and under the free edge

Step 14

Complete cuticle work on left hand

Step 15

Remove right hand from the bowl and pat dry

Step 16

Apply cuticle remover to cuticles of right hand – fingers 6–10 – under the free edge

Step 17

Complete cuticle work on right hand

Step 18

Brush the nails of both hands over the bowl

Step 19

Inspect both hands for any rough edges or remaining loose cuticle

Step 20

Remedy, if necessary, before proceeding

Step 21

Hand massage *both* hands

Step 22

Clean nail plates of both hands with nail enamel remover

Step 23

Apply base coat to fingers 1–5

Step 24	Apply base coat to fingers 6–10	
Step 25		Apply first coat of enamel to fingers 1–5
Step 26	Apply first coat of enamel to fingers 6–10	
Step 27		Apply second coat of enamel to fingers 1–5
Step 28	Apply second coat of enamel to fingers 6–10	
Step 29		Apply top coat (if using cream enamel) or third coat (opaline polish) to fingers 1–5
Step 30	Apply top coat (if using cream enamel) or third coat (opaline polish) to fingers 6–10	
Step 31	Both hands – quick-dry spray, if required	

Unit 4

Manicure Techniques

REMOVAL OF POLISH

The main ingredients of polish remover may be acetone or amyl acetate or ethyl acetate, and materials such as castor oil, or more recently butyl stearate or dibutyl phthalate.

There are two distinct types of remover:

1 For removing acetate-based polishes
2 For removing plastic or nylon-type polishes.

The remover for one type of polish does not easily remove the other type, so the remover should be chosen according to the polish used. Familiarity with manufacturers' products will indicate which is the correct one to use on each occasion. An acetone-free remover is more gentle on the nail plate and will not dry the nails quite so much.

It is important that the manicurist's nails do not get smudged while old polish is being removed from the client's nails. For this reason it is advisable to get used to holding cottonwool between the index and second finger at the first joint below the nail plate.

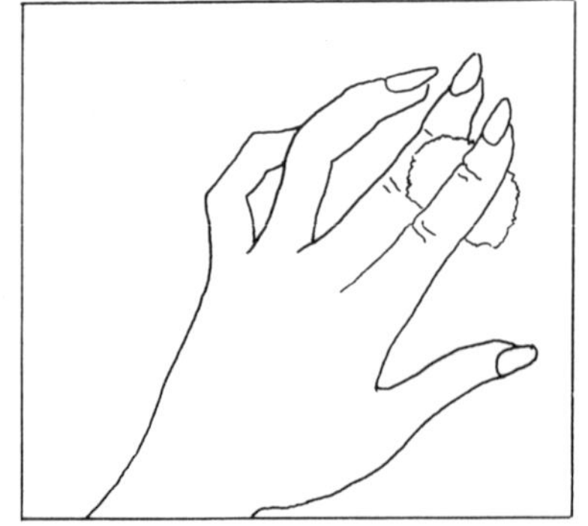

Polish removal – hold cottonwool at first joint

First Stage Removal

Apply nail polish remover generously to a piece of cottonwool that has been rolled first in the palms of the hands to reduce 'fuzz'. Then press firmly on to the nail and allow the old varnish to soften. The manicurist will be able to feel the enamel begin to dissolve. Dark shades, or several coats of enamel, will take longer to soften.

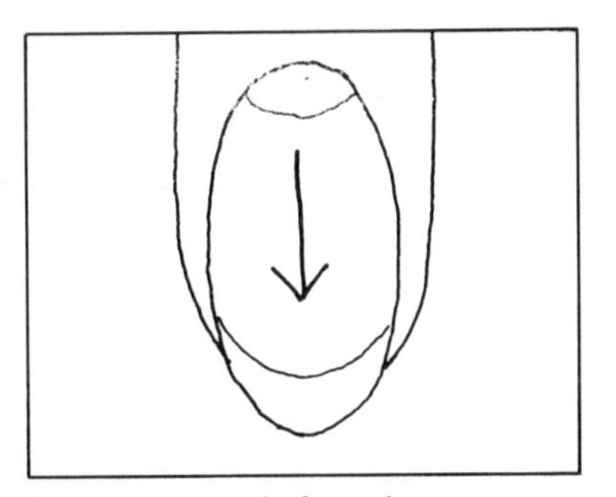

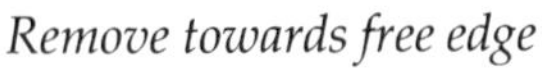

Remove towards free edge

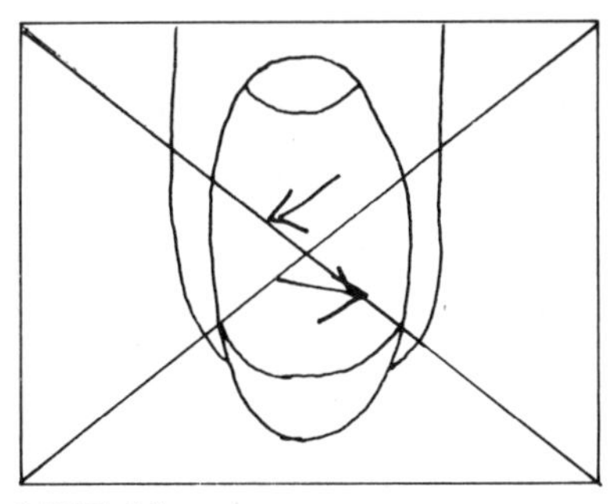

NOT this way

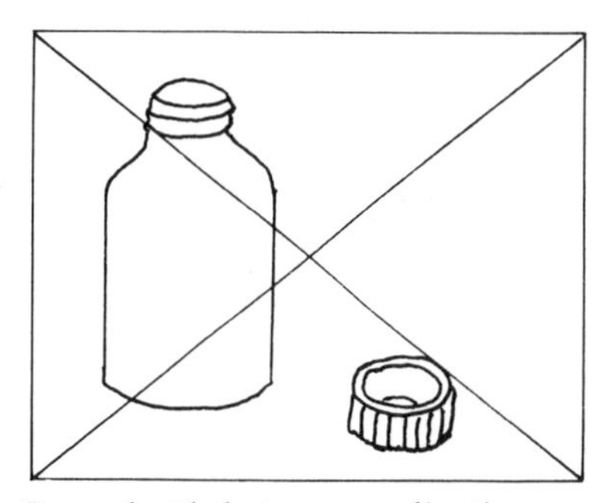

Recap bottle between applications

The direction of removal must be from the cuticle towards the free edge. Rubbing the enamel in a see-saw manner will smear the polish under the cuticle where it becomes very difficult to remove.

Fresh cottonwool may be needed for each nail but this is a decision only the manicurist can make during the removal procedure. It is not likely that all the polish will be removed from around the cuticle by using the first stage but resist the temptation to rub polish into the cuticle.

The same first-stage procedure should be followed for both hands.

Do not forget to replace the cap of the polish remover bottle between applications to the cottonwool. If the bottle is accidentally knocked over, spilt polish will dissolve the surface of plastic tables, trolleys, floors, etc. It can also cause extensive (and expensive) damage to synthetic materials in clothes and carpets. It is highly inflammable and the fumes are toxic.

> **Safety Tip**
>
> Special care is needed when disposing of cottonwool soaked with nail polish remover because of its inflammability and toxicity. It must be placed into a sealed bag and disposed of separately – immediately – into external bins.*

Second Stage Polish Removal

If there is still polish around the cuticle remove it using the

*COSHH regulations – Control of Substances Hazardous to Health.

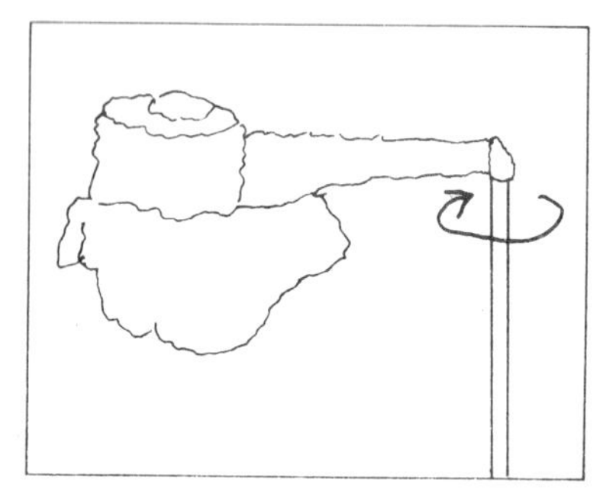

Roll orange stick to wind cottonwool

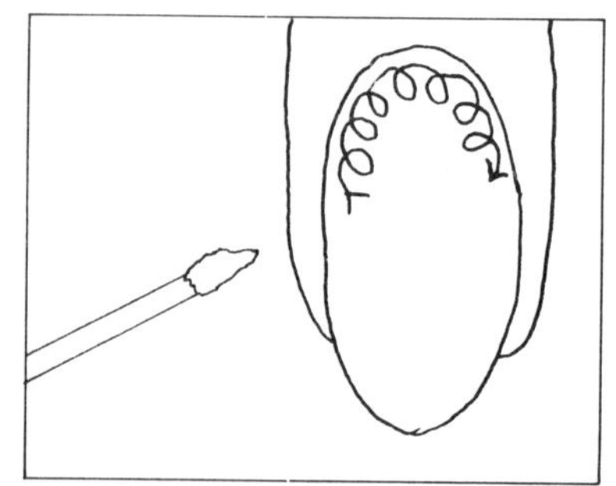

Clean around cuticle with circular motions

Place orange sticks in a jar containing 70% alcohol when not in use.

wedge end of an orange stick wrapped in cottonwool. The most effective way to wrap the stick is by holding the stick horizontal to the cottonwool and rolling it until enough cottonwool adheres. The making of candy floss gives the general idea except that the cottonwool should be wound tightly round the stick, not made into a fluffy bundle. The manicurist should still be able to feel the wedge of the orange stick.

The grade of cottonwool used is important. It is difficult to achieve an even, firmly packed cottonwool stick with low-quality mixed-fibre wool. It is easier with high-grade, surgical, absorbent cottonwool. But the better quality cottonwool is more expensive which must be considered when costing manicure services. (See Unit 17)

Dip the cottonwool-tipped orange stick into remover and carefully remove any remaining polish around the cuticle area. Use a circular movement, changing the cottonwool when necessary. When all the polish is completely removed proceed to the next step of shaping the nails.

SHAPING THE NAILS

The intended shape of the nails must be fully considered. A number of questions need to be asked and a variety of factors will influence the final decision. These are as follows:

1 What are the client's wishes and requirements?
  It is important to discuss the needs of the client and agree what is to be achieved.

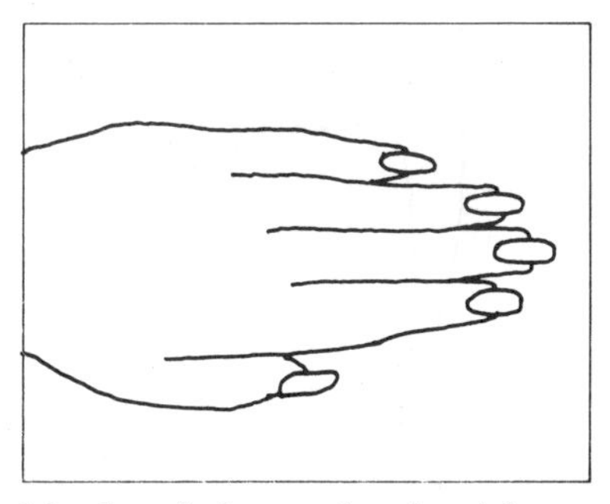

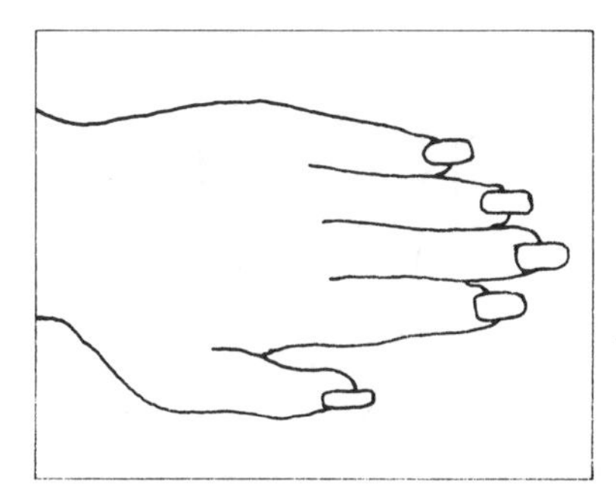

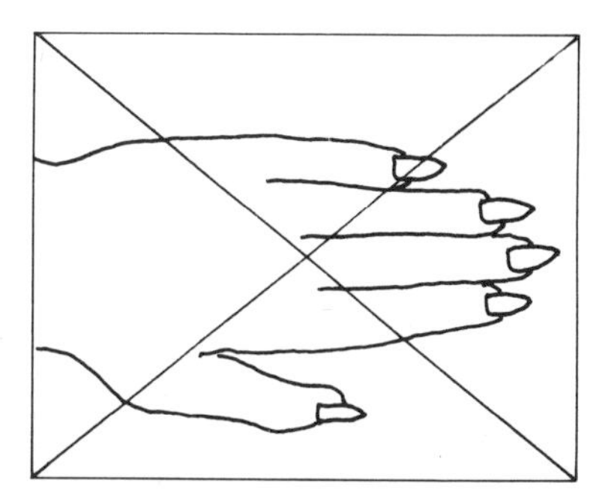

Match nail shape to hand and finger shape

2 What limitations are imposed by the present state and shape of the nails?
If these are ignored the finished effects can be badly marred and excuses will need to be made.

3 What is the condition of the nail plate?
Assess the thickness, texture, dryness, damage, colour and blood circulation.

4 Is the client's lifestyle and occupation a significant factor in deciding final shape and length?
It is not usually practical to encourage long, tapered nails for people who, for example, work in the garden all day. But a model may display the longest tapered nails.

5 What is the shape of the hand?
A slender hand with tapering fingers will be enhanced by long tapering nails. Broad, clubbed fingers will look most becoming with a free edge that extends only slightly beyond the fingers.

6 What is the present shape of the nails?
Nails can be oval, square, broad, small, large, wide or narrow. To achieve balance the shape of the nail at the tip looks best if it reflects the shape at the base.

7 Are all the nails equal in shape and length?
Nails look better if they are of similar shape and length. If

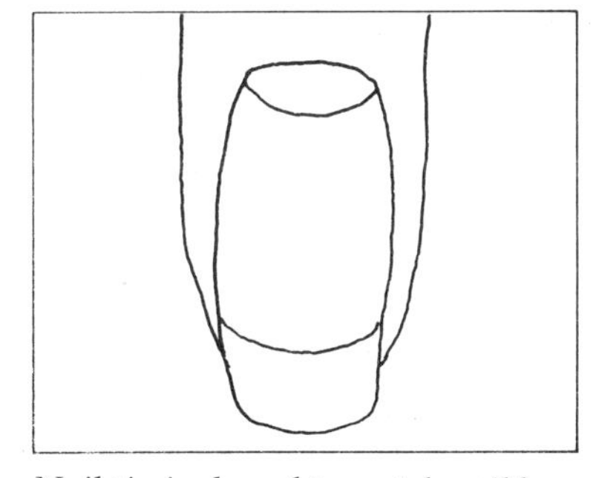

Nail tip is shaped to match nail base

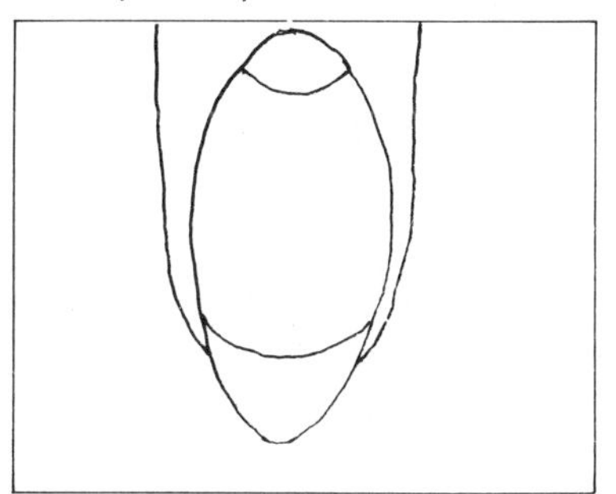

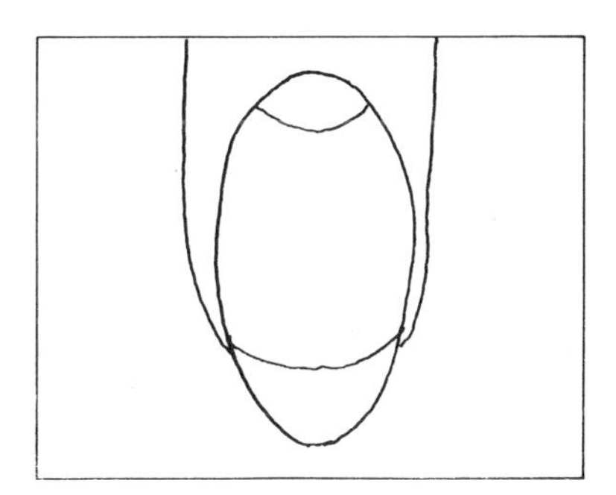

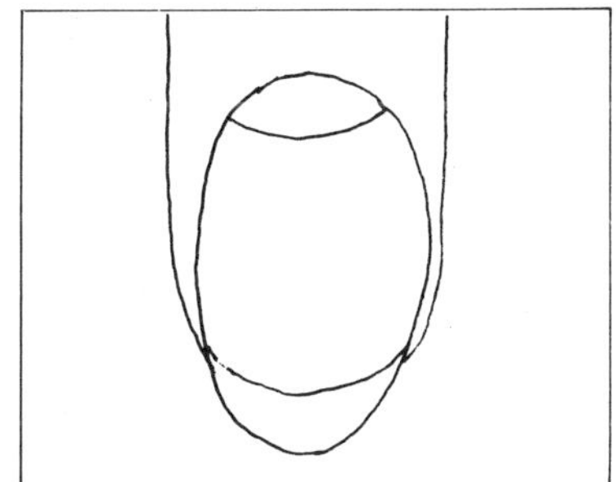

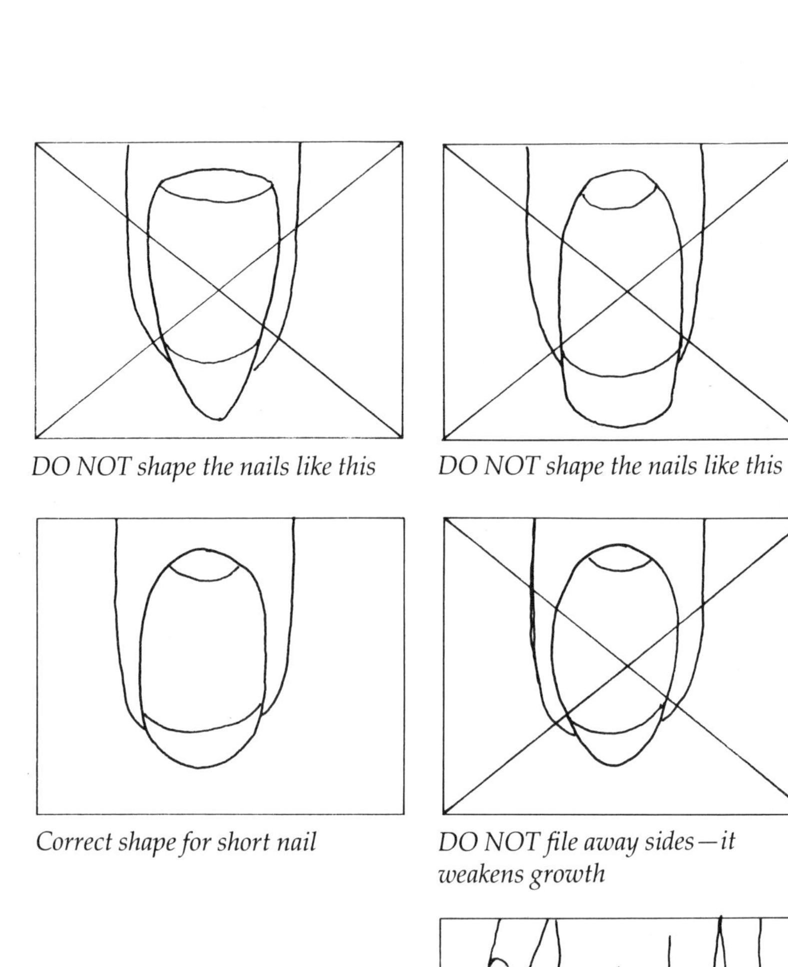

DO NOT shape the nails like this

DO NOT shape the nails like this

Correct shape for short nail

DO NOT file away sides—it weakens growth

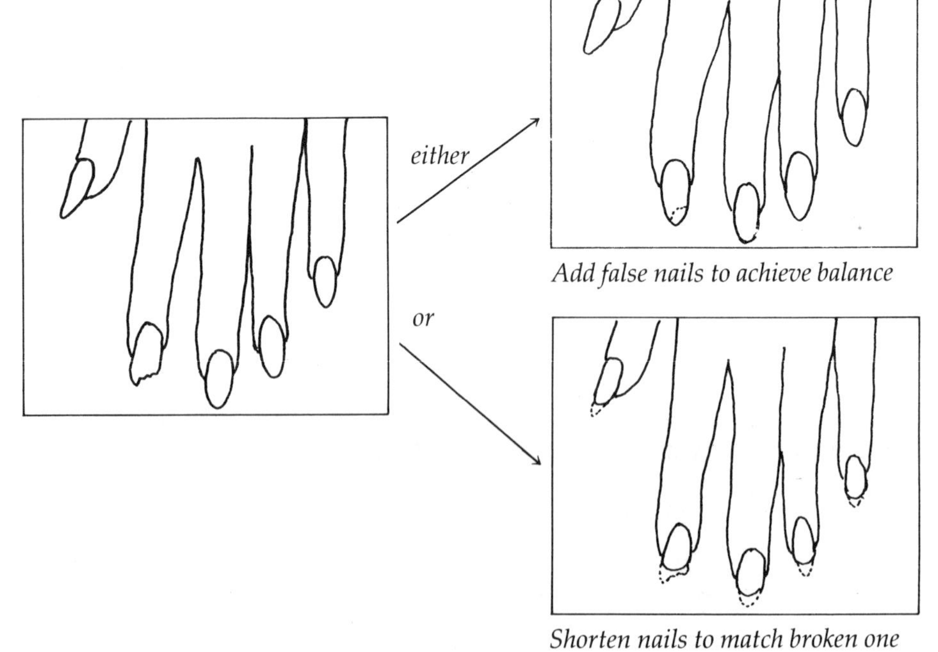

Add false nails to achieve balance

Shorten nails to match broken one

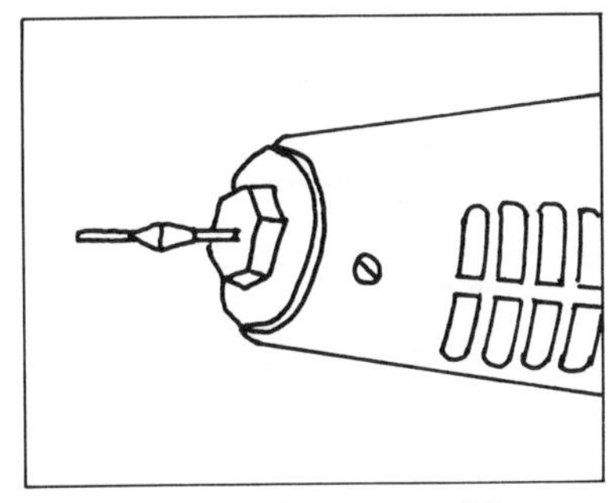

Electrical nail shaping machine

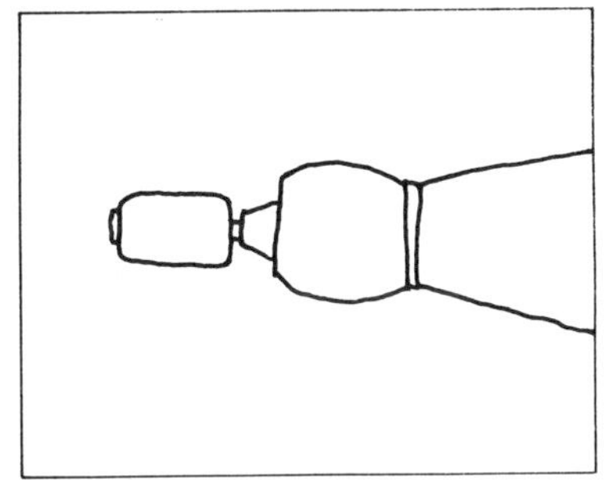

Battery-operated filing system

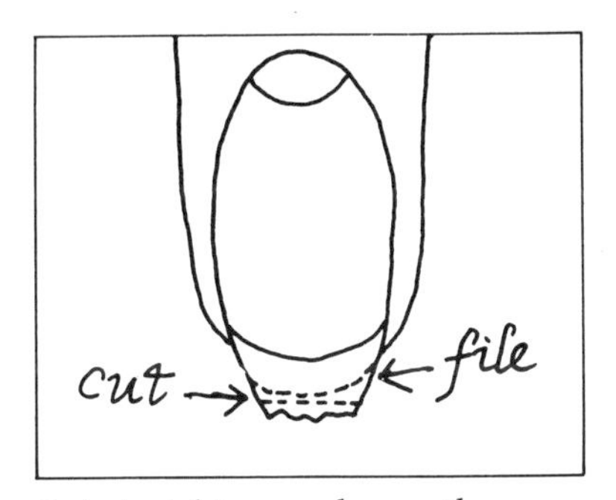

Cut straight across longer than final shape

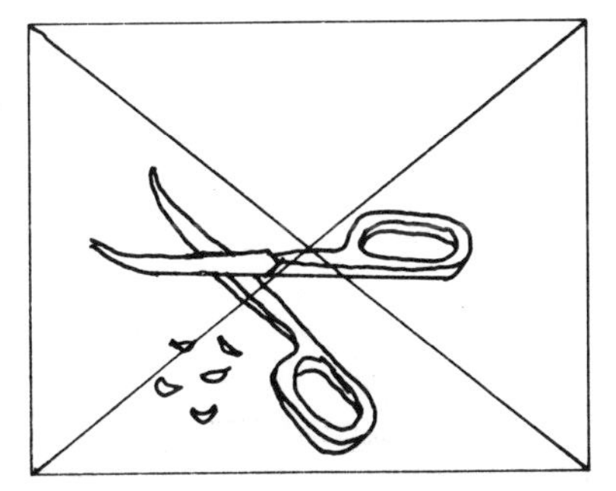

Dispose of cuttings, sterilize scissors

one nail is short or broken it may be best to shorten the other nails or consider the addition of a false nail to achieve balance. If the nails are breaking then find out why.

When all these questions and factors have been considered, discussed and agreed then the shaping techniques can be embarked upon.

Electrical machines have been developed to help shape nails. They are usually used only for filing nail extensions, and their only advantage is that they are slightly faster. The same principles apply as when using emery boards. The nails will usually require finishing touches with an emery board. Battery-operated systems for filing are becoming more popular for home manicures.

Cutting the Nails

Cutting the nails is not normally recommended as it encourages the nail plate to layer or separate. Nails must only be cut if the length is to be reduced substantially and the nail plate is so strong that a long time would have to be spent on filing.

If nails are to be cut, make sure the scissors are small and comfortable to use. They must be sharp to avoid breakage and move smoothly for easy use. They must be clean, and sterilized by wiping with alcohol before and after use. Scissors must be kept in good working order by oiling with light machine oil or sterilizing oil and must be sharpened professionally as required.

The nail should be cut straight across, leaving it slightly longer than the desired final length, without compromising the final shape.

Safety Tip

Take care to dispose of all nail cuttings in a closed bin and sterilize the scissors by again wiping with alcohol before placing into an ultraviolet light disinfecting cabinet where they can be safely stored until needed again. They should be autoclaved if they have been in contact with cuts or open wounds (see p. 119).

Filing the Nails

The nails are shaped using emery boards. These are strips of cardboard coated with emery, i.e. corundum (a natural aluminium oxide) and iron oxide, which provides an abrasive surface for paring down the nails. Emery boards must be flexible for the best results and the cost is a factor that must be balanced against quality. Those of finer texture and greater flexibility are more expensive. It is useful to have a supply of cheap emery boards, which are larger, coarser-textured and less flexible to use for preliminary heavy-duty filing where a lot needs to be removed from a particular nail.

Emery boards are usually made with one end broader than the other. Hold the board at the broad end with the fingers on one side and the thumb on the other. If removing considerable length from thicker nails use the coarser side first. Always finish with the finer side to ensure smoothness.

Incorrect filing – rough, hard, with a sawing action – produces friction and causes the nails to heat and split into layers. Clients should be advised on the correct method of filing to use when caring for their nails.

A new emery board should be used for each client. If there is any life in the board when the manicure is finished, offer it to the client for use at home. If the client does not want it, snap it and dispose of in a covered bin. This helps avoid the spread of infection.

Nails must be filed from the sides to the centre, not back and forth which causes layering. The emery board should follow a smooth, rounded path at an angle of 45° under the nail, to achieve the desired shape. The pressure should be upwards towards the free edge and each stroke should use

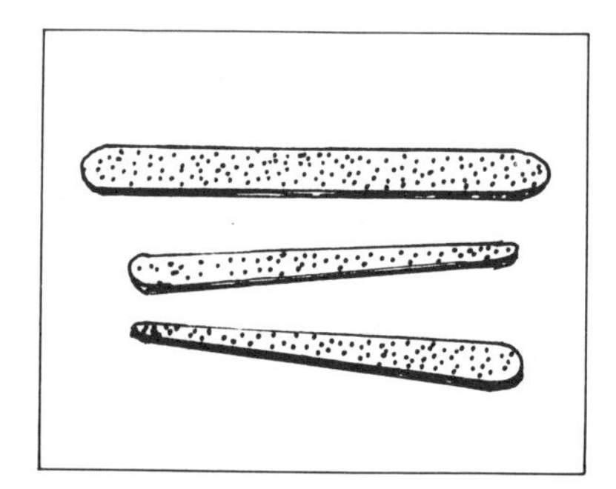

Emery boards

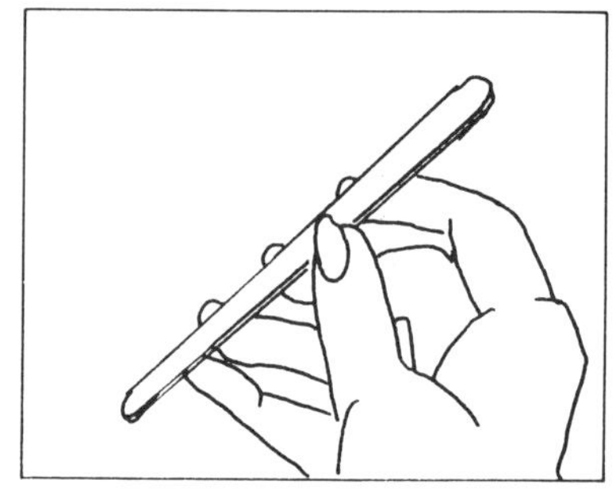

Holding an emery board

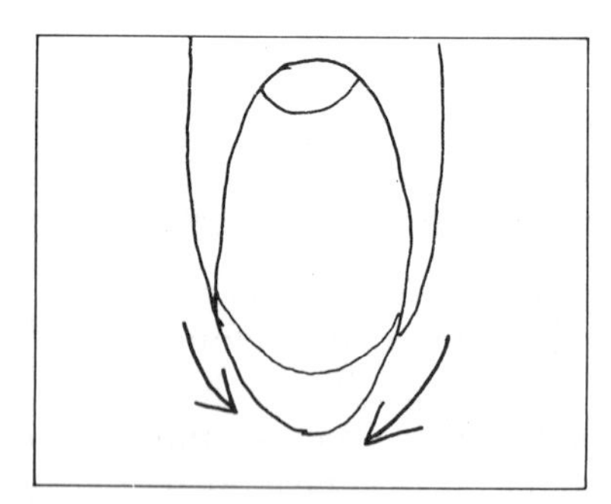

File from sides to centre

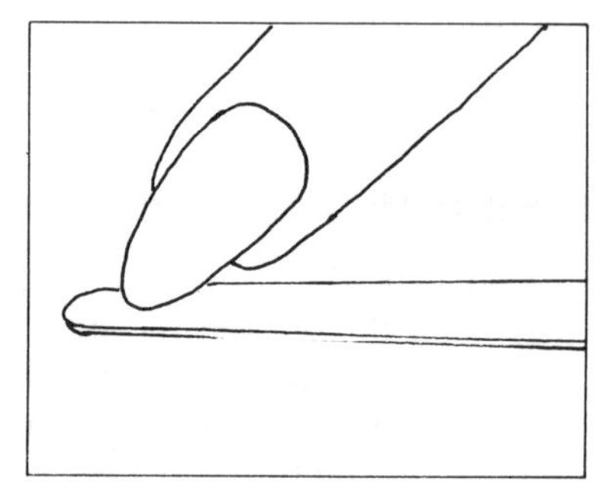

Maintain angle of 45° to nail

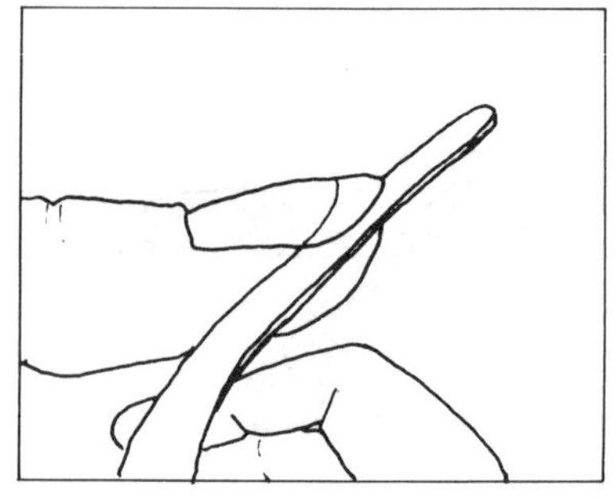

Use full length and flexibility of emery board

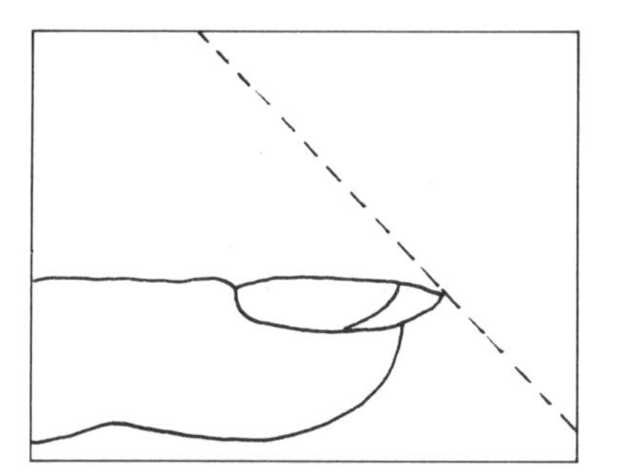
Bevel with fine side of emery board at 45° angle

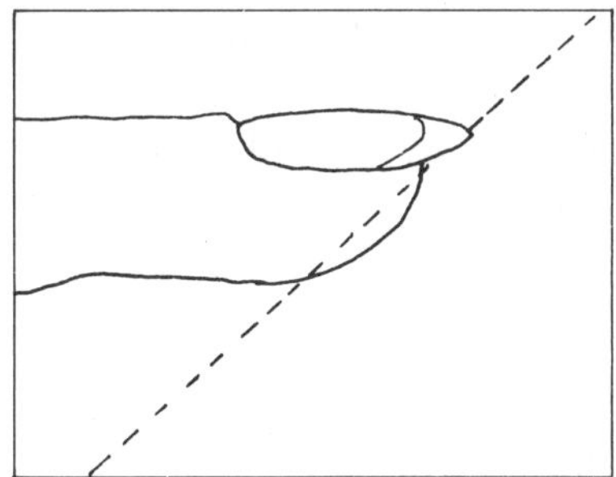
Bevel from below also at 45° angle

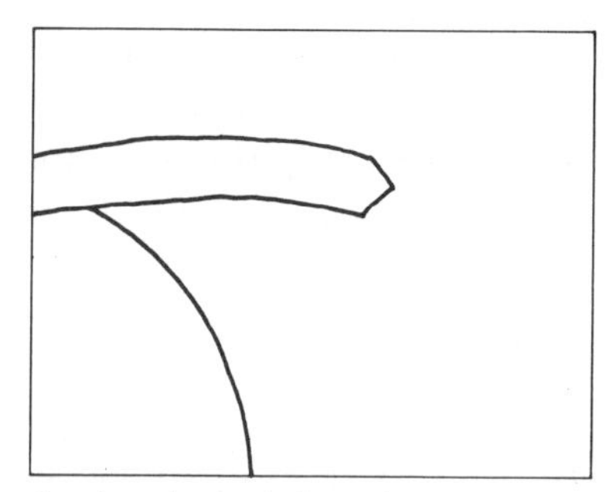
Section through free edge (exaggerated scale) showing bevel

the full length and flexibility of the emery board for maximum effect. Take care that the back of the board does not friction-burn the client's skin. Check frequently that discomfort is not being caused.

Do not file into the corners of the nails as this can cause ingrowing nails and weaken the sides of the nails. Do not file the sides of the nail away because, as well as looking unpleasant, this lessens the support to the free edge and leads to breakage. The nails can be carefully bevelled with the fine side of the emery board to prevent layering. Follow the procedure carefully taking each nail in turn.

Buffing

A chamois-covered buffer

Buffing is often used when the client does not want nail polish, or in men's manicure. Several occupations preclude the use of nail enamel, such as nursing and catering. If this is the case then buffing will be used at the end of the manicure.

Buffing is carried out with a leather-covered buffer, and a buffing paste which usually contains powdered pumice rock, tin oxide and chalk. The paste is not always necessary.

Buffing is beneficial in the following ways:

- It improves and stimulates circulation to the nail matrix and the nail bed – with or without buffing paste.
- It gives the nail plate a pink, healthy, natural shine.
- It minimises ridges in the nail plate and allows a smoother application of polish, particularly with the use of buffing paste.
- It is claimed that additives such as glycerol, cholesterol

and vitamin A in the buffing paste can strengthen and improve the nails.

- It helps stop the nails splitting into layers and improves the appearance if there is any layering.

The Buffing Technique

Buffing paste should be removed from the pot with an orange stick. Never use fingers as this could contaminate the whole container and pass infections from client to client. A small amount of paste should be touched on to the nail just above the half moon. Take care not to use too much paste – it will find its way under the cuticle and be difficult to remove. An amount less than the size of a grain of rice is the maximum needed for one nail. Spread the paste with the ball of the thumb, towards the free edge, before using the buffer.

Buffing paste is a mild abrasive and is used for reducing minor nail ridges, so it should not be used when the nails are smooth. It can be detrimental to use buffing paste when there are deep ridges in the nails.

The buffer should be held between the fingers rather than between the thumb and finger. This stops too much pressure being applied which could cause friction. The heat of friction is uncomfortable for the client and can damage the matrix. Each stroke is from the cuticle to the free edge. The buffing action must not be too fast and the buffer should be removed from the nail after each stroke. This too will avoid excessive friction. Try the buffer on your own hand and see just how hot it gets if it is used too fast with too much pressure. Check frequently with client that the heat generated is not uncomfortable.

Remove paste from pot with orange stick

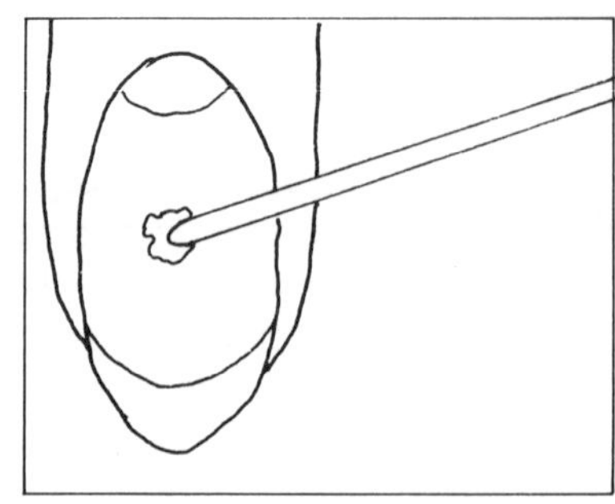

Apply buffing paste above half moon

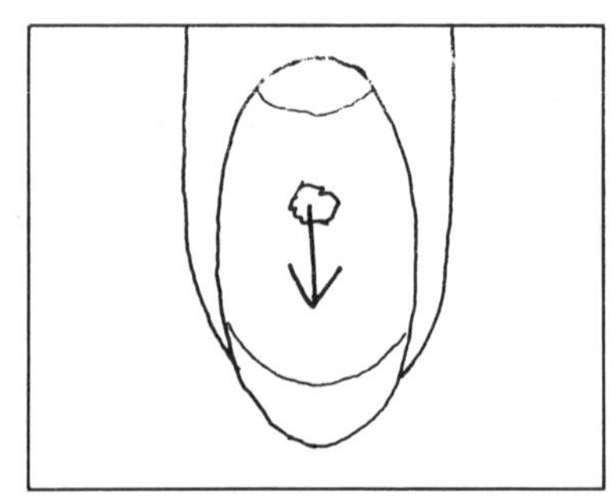

Spread with ball of thumb towards free edge

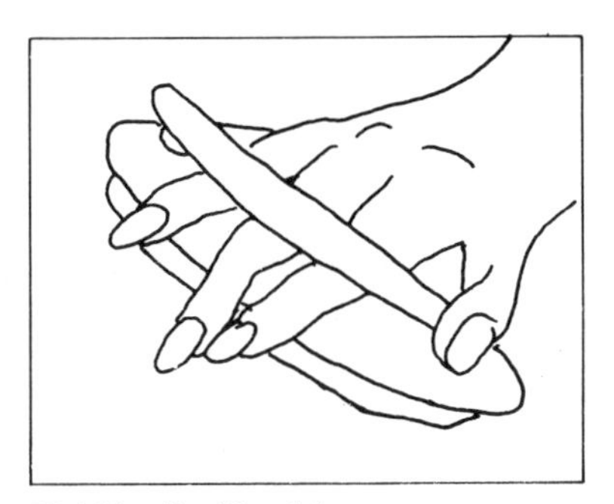
Hold buffer like this

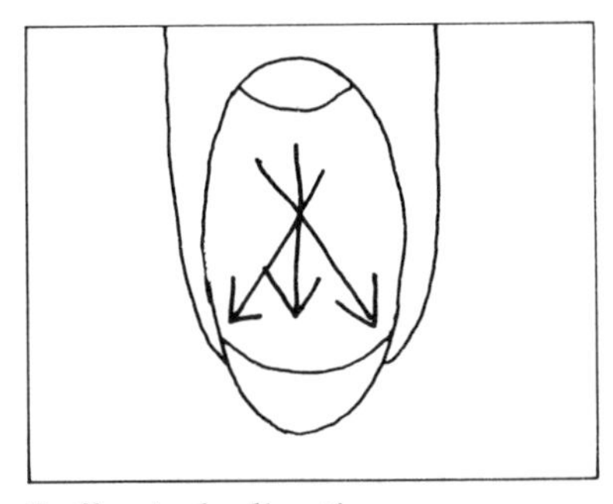
Buffer stroke directions

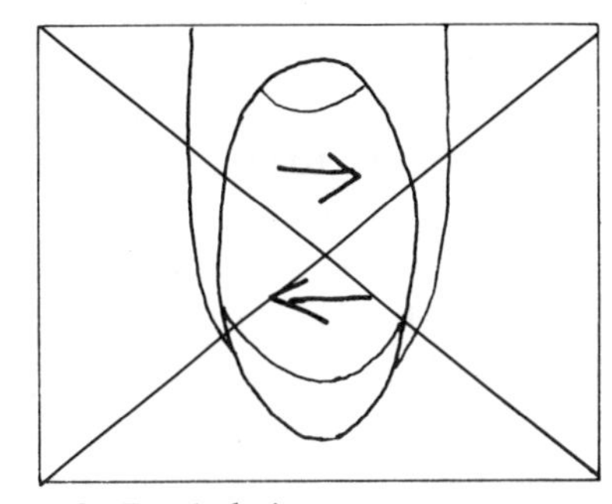
NO! Don't do it

The direction of buffing must never be across the nail plate. Use diagonal and vertical strokes as this will most easily buff the complete surface of the nail plate.

CUTICLE TREATMENTS

Cuticle Massage

Cuticles are massaged with a rich emollient treatment cream containing lanolin, glycerol, ethylene glycol or sorbitol, and white iodine. This has the effect of softening, enriching and conditioning the cuticles.

The cream is taken from the pot with either a spatula or an orange stick. It can be applied either directly to the cuticles or placed on the back of the manicurist's hand. A generous amount can be used but do not dispense too much as this can be wasteful.

The cream should be well rubbed in, using the balls of the thumbs in small circular movements around the cuticles. Time can be saved by using both left and right hands together on different fingers. Particular attention should be given to the index finger and thumb as these cuticles are likely to be in the poorest condition because of the wear and use these digits are subjected to, especially on the hand the client uses most.

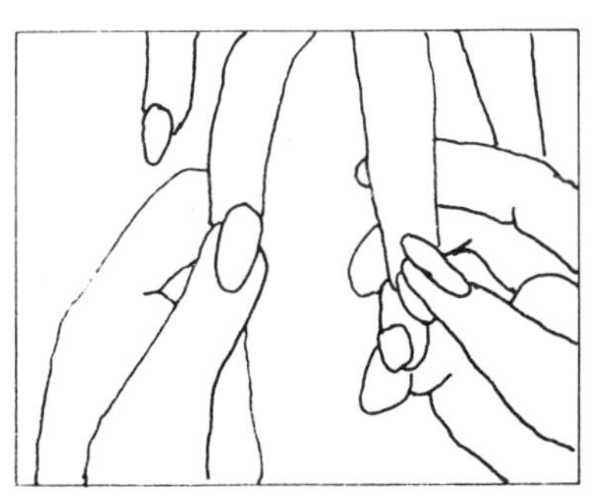
Use both hands to save time

Soaking the Cuticles

After massage, the cuticles will need soaking in warm, soapy water. This softens and loosens the cuticles in preparation for further treatment. Do not get the water before it is required or it will grow cold. Manicure 'pills' are available as an alternative to soap. They are effervescent and have a mild bleaching effect.

The bowl containing the water must be placed in a position that is comfortable for the client and which does not stop the manicurist continuing treatment on the other hand. There must be no risk of spilling.

Leave the first hand to soak while completing filing, buffing and cuticle massage on the second hand. When this is completed remove the first hand from the water and pat dry gently with a towel. Test the temperature of the water. If it has gone cool replace it before immersing the second hand. Use the towel to dry gently around the cuticle, pushing it back at the same time. Treat each finger separately.

With experience, the procedure may be altered so that one hand is given longer to soak than the other. The hand that the client uses most tends to need more soaking.

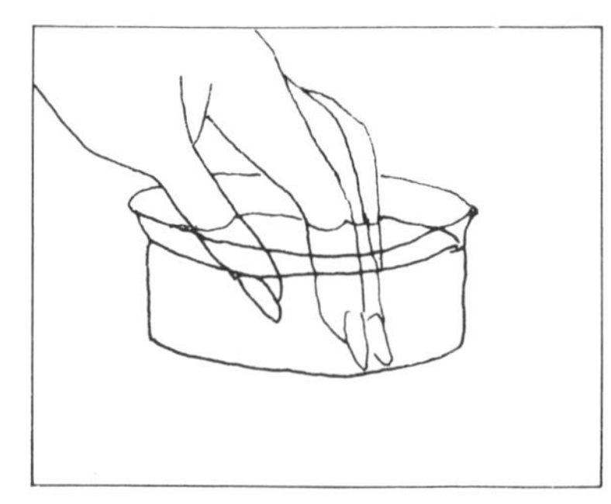

Soak nails in soapy water

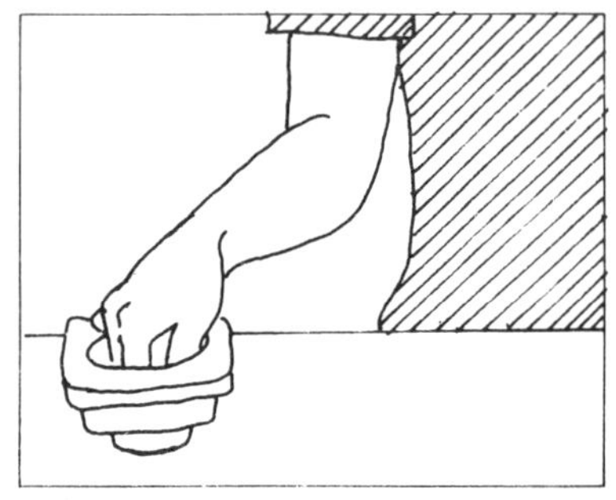

Place bowl comfortably for client but safe from spilling

Warm Oil Treatments

Warm oil treatment is useful for hands and nails that have been badly neglected and need intensive care. It is often required by people preparing for a big event such as a wedding. The oil – olive or almond – helps replace natural oils removed through harsh treatment and the effects of detergents. It makes the subsequent cuticle treatment easier for the manicurist. Oil treatment is particularly helpful when dealing with conditions such as pterygium and onychorrhexis (see p. 110).

If applying a warm oil treatment, substitute a penetrating cuticle oil for cuticle massage cream, and massage in the same way. Then use warm oil instead of hot soapy water.

Oil heaters are available which keep the oil at a constant temperature. A thermostat makes sure the oil does not overheat as this could be dangerous.

The heater should be switched on at least ten minutes before it is needed. The manicurist must check the temperature before the client's hand is immersed. How long the hand is soaked in the oil depends on the condition, 10–15 minutes is recommended. After taking the hand out of the oil repeat the massage procedure, then gently remove the oil

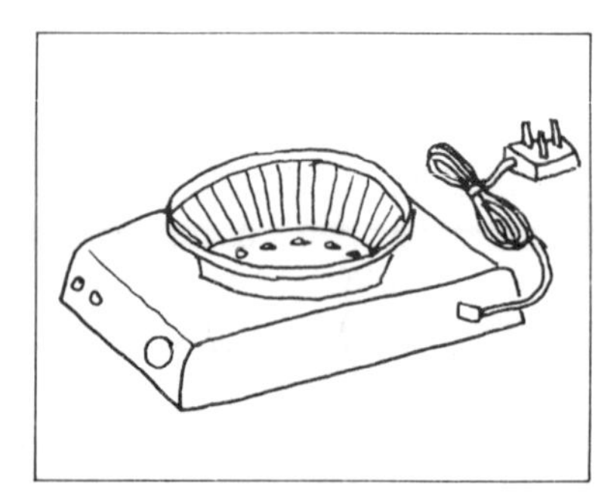

An oil heater. Switch on 10 minutes before use

with a tissue. Proceed with the cuticle treatment, but in cases of extreme dryness omit the use of cuticle remover as this could cause further damage.

Clients in need of warm oil treatments will need advice on causes of the damage. Weekly treatments should be recommended and the client should be advised on the need for home care between treatments. Paraffin wax treatments for the hands might be discussed. If the nails and cuticles are badly dehydrated then the hands are likely to be in a similar condition.

Home Care

The client can buy some oil from the manicurist and heat it at home in a glass standing in hot water. This will not be as effective as using an oil heater because it is impossible to keep the temperature constant. The oil must never be heated directly because of the dangers of overheating and fire. Penetrating cuticle oil can also be suggested and daily use will soon produce an improvement.

Cuticle Remover

Cuticle remover (solvent) is used to loosen and remove dead cuticle from the nail plate. It has a mild bleaching action, which removes stains from the nail plate. It also helps to remove debris and stains from beneath the free edge.

The cuticle remover often contains a strong alkali, such as potassium hydroxide, combined with a variety of additives, such as glycerine or propylene glycol, which are included to reduce the drying effects of the alkali. Cuticle softeners based on quaternary ammonium compounds are also used.

Nearly all brands of cuticle remover or softener are supplied with a brush for application. But for reasons of hygiene it is better to dispense a small amount of remover into a container the size of a small bottle cap and then apply it with a cottonwool-tipped orange stick or a small brush that can be sterilized after use.

Cuticle remover should be applied all around the cuticle and under the free edge. This allows the cuticle to be loosened from the nail plate easily and removes any dead skin beneath the free edge at the flesh line.

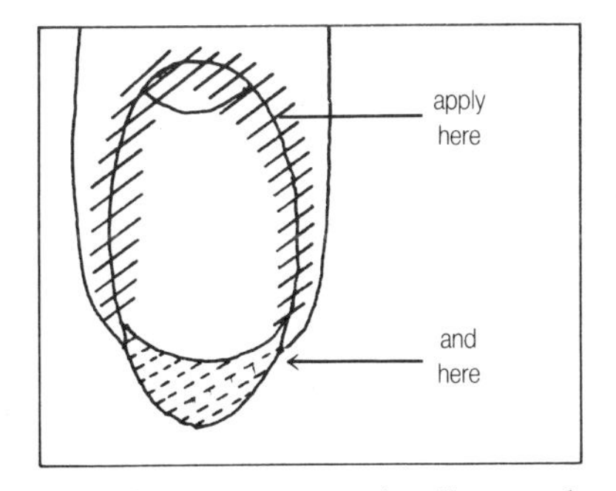

Cuticle remover—apply all around cuticle and under free edge

Start with the little finger. Apply enough to keep the area moist while working. Avoid fingers with hangnails or open cuts as the alkali in the solvent will sting and cause discomfort to the client.

After the application of the cuticle remover/softener/ solvent the cuticle can be removed as follows:

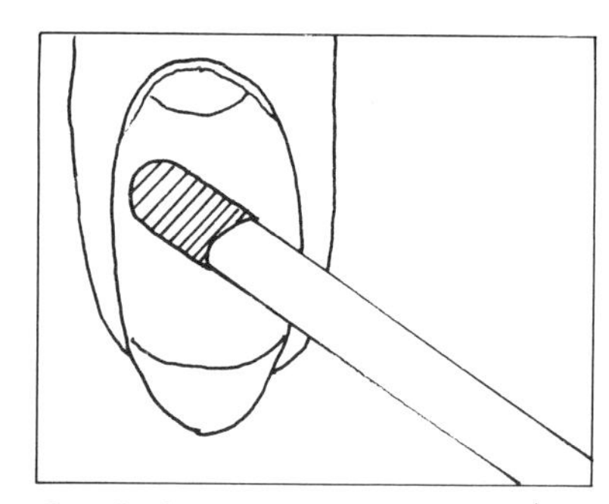
Gently free cuticle from nail plate

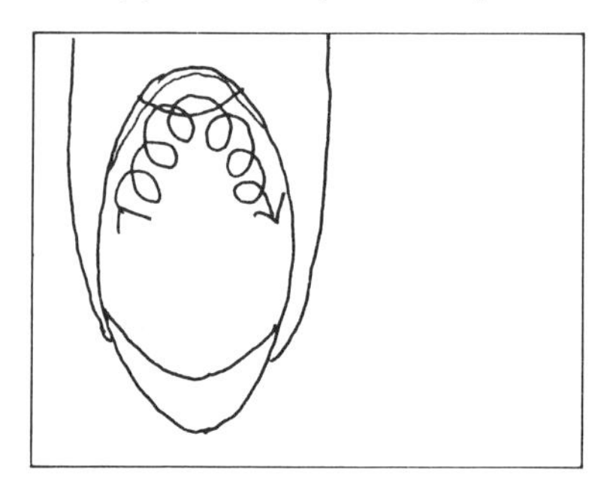
Use circular motion

1 Use the wedge end of a cottonwool-tipped orange stick or a rubber-ended hoof stick to loosen the cuticle. Gently free the cuticle from the nail plate with circular movements all around the cuticle. This must be done carefully or the cuticle will tear causing hangnails. If too much pressure is applied to the matrix it will be damaged and horizontal ridges will form in the nail plate and emerge as the nail grows.
2 If any cuticle is still adhering to the nail plate it may be necessary to use the cuticle knife. This must be sterilized with alcohol or autoclaved before use. The cuticle knife is not likely to be necessary for a client who has regular manicures. Great care must be taken not to scratch the nail plate, cut the cuticle, or exert undue pressure on the matrix. The knife should be held parallel to the nail plate and used in small circular movements to scrape away gently any remaining cuticle. Do not dig under the cuticle.
3 Remove excess cuticle after scrubbing the nails.
4 Cleanse the underneath of the free edge of the nail using the cottonwool-tipped, pointed end of the orange stick. Use a rolling movement of the orange stick to remove dirt and debris. Do not be tempted to dig or poke with the orange stick as this will be uncomfortable for the client and could introduce infection to the nail bed.

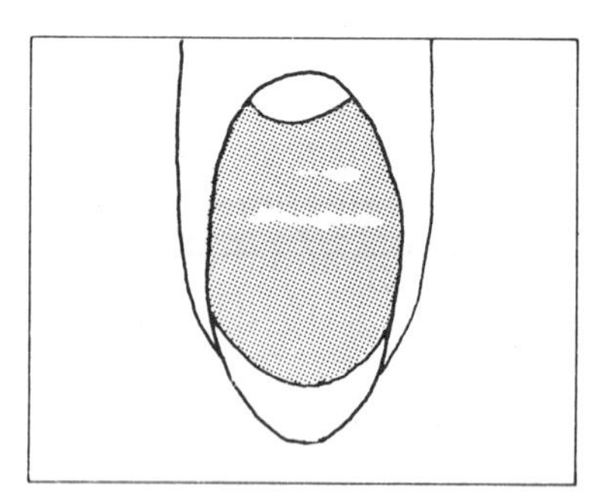
Ridges resulting from pressure damage to matrix

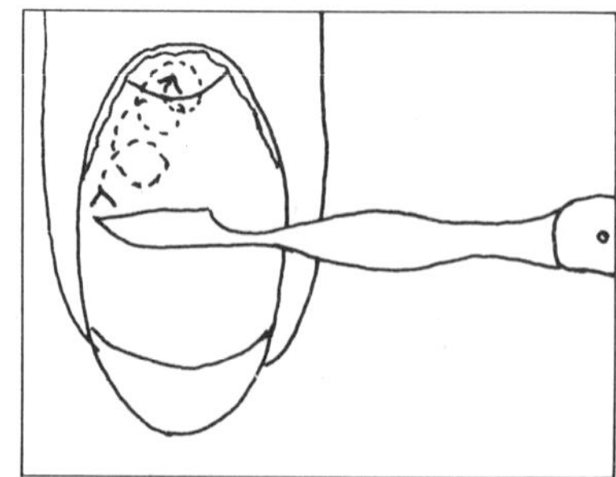
Scrape away cuticle in circular movements from sides to centre

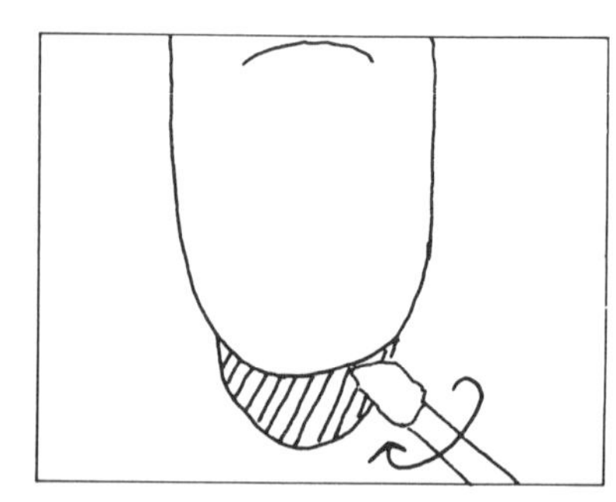
Clean underneath using rolling motion

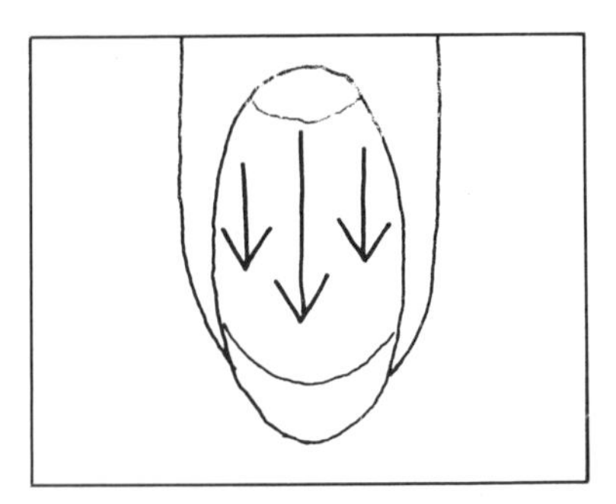

Brush this way

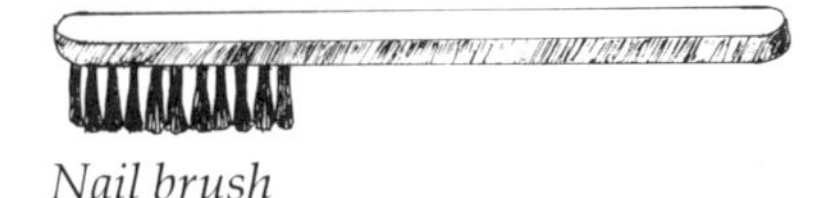

Nail brush

When cuticle removal has been completed for all the fingers of one hand, remove the other hand from the water and repeat the cuticle removal procedure.

Brushing the Nails

The purpose of brushing the nails is to remove the remains of cuticle solvent, dead cells around the cuticle, and any debris under the free edge. It should also remove traces of creams, oils, or solvents used up to this stage of the procedure.

The nails should be gently brushed with soapy water in a downward direction from the cuticle towards the free edge. The choice of brush is important. Most nail brushes are too large to be useful for this purpose and the bristles too hard to be comfortable for the client. A very stiff nylon bristle brush will also tend to flick splashes on to the client, manicurist and work area. Ideally, the brush should be slightly larger than a toothbrush, with natural bristles rather than harsh nylon tufts. It must be able to withstand sterilization with disinfectants (see p. 118).

When the brushing is complete, gently dry the hands, fingers and nails ready for inspection.

NAIL AND CUTICLE INSPECTION

It is at this point that the effects of treatment must be evaluated and any faults corrected. The inspection should be carried out as follows:

1 Check to see if the filing has achieved the correct and uniform shape and length. Make corrections as necessary.
2 Make sure the free edges of the nails are smooth, with no evidence of separation of the layers. Don't just check visually. Run your finger around the free edge to check for any roughness that may cause the client to 'snag' the nails later. Correct by bevelling if necessary (see p. 23).
3 Check that the cuticles are free of the nail plate and that there is no excess cuticle. Correct if necessary by repeating the cuticle removal procedure thoroughly.

Advise the client with neglected nails that regular manicure will give a better result each time. Each cuticle treatment will reveal a fraction more lunula, and correctly filed and nourished nails will lengthen and strengthen.

4 Correct any rough, torn, or overhanging cuticle present by using the cuticle nippers (see below).

5 When all these checks have been made clean the nail plate with polish remover to remove any traces of oil or other products. Then proceed to the base coat polish application, if required (see p. 32).

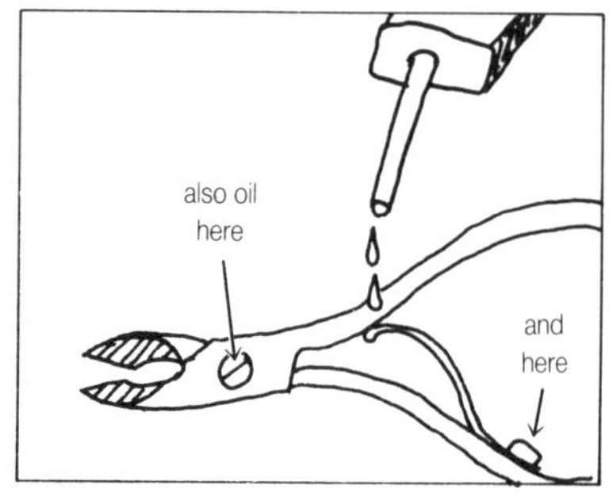

Lubricate spring

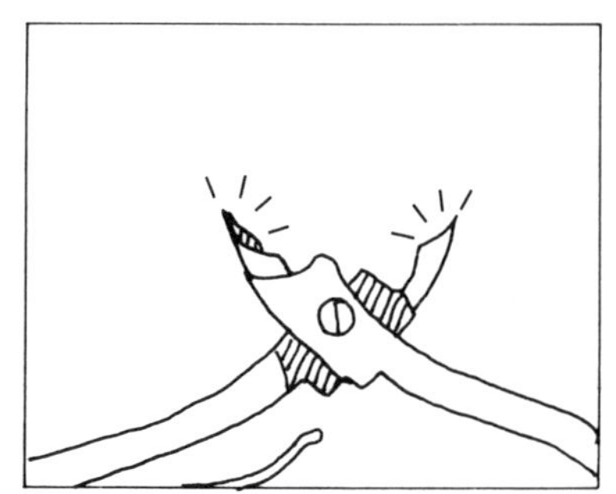
Check blade sharpness often, but let a professional do the sharpening

Cuticle Nippers

When selecting cuticle nippers make sure they are the right size to fit snugly into the hand. They should have a smooth movement and a durable spring. The cutting edge should be small to ensure accuracy and they must be comfortable to use. Avoid cheaper types of cuticle nippers which may have inferior metal, spring and cutting edge, which could cause the cuticle to be torn rather than cut.

Cuticle nippers must be cleaned and sterilized with alcohol before and after use. And if they come into contact with blood they should be autoclaved. After sterilization, store them in an ultraviolet ray cabinet.

Nippers are best stored in a case to protect the blades. If the edges become dull they must be resharpened. The spring must always be kept lubricated with fine machine oil.

Use of Cuticle Nippers

The cuticle nippers should be held firmly in the palm of the hand. The nipper edges should be placed just under the portion of the cuticle to be removed. The blades should then be closed decisively to cut the cuticle firmly. Great care must be taken to avoid sliding the blades too far under the cuticle or the dermis will be cut, causing bleeding. If this should happen, immediately apply cold, wet cottonwool and slight pressure until the bleeding stops. Then apply styptic containing alum. If a styptic stick is used apply it first to cottonwool rather than directly to the wound, or make sure

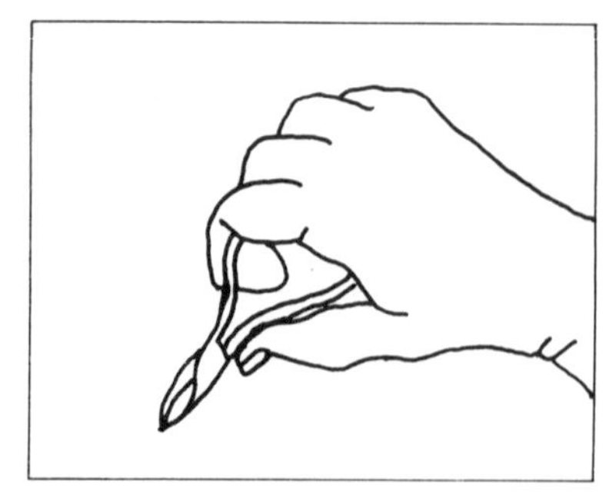
Hold cuticle nippers like this

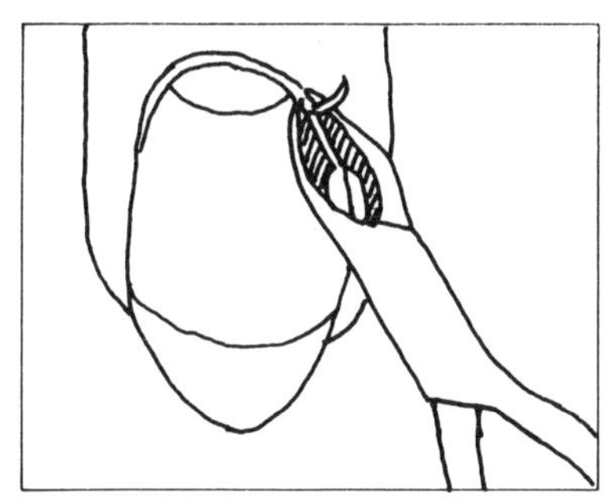

Remove cuticle in one piece

that it is disposed of after use to prevent cross-infection. Alternatively powdered alum may be used.

It is important that all the cuticle is removed in a single piece so that there are no rough edges which might tear and cause hang nails. Do not be tempted to cut the cuticle away completely as this will weaken it and again cause hang nails. This can also lead to bacterial infection and felon/whitlow/paronychia can result (see p. 111).

Cuticle nippers can be extremely effective for rapidly improving the appearance of excessive forward cuticle growth (pterygium). Used in conjunction with oil manicure nippers can eradicate the condition.

Cuticle nippers must be used carefully when removing torn cuticle causing a hang nail (see p. 104), as they can easily make the condition worse and cause discomfort to the client.

Trainee manicurists should familiarise themselves with the use and effect of cuticle nippers by practising, on themselves at first, until they gain confidence.

POLISH APPLICATION

It is usual to complete a hand massage before applying polish (see Unit 5). In this way the polish application is left until last and doesn't become smudged.

Preparation

1 Inspect the nail and cuticle thoroughly (see p. 29).
2 Check to see if patching is required (see p. 46).
3 Clean nail plates with polish remover to eliminate all traces of oil or cream – particularly massage cream. This step is essential for an even, hardwearing polish application. Any grease left on the nail plate will prevent the adhesion of the polish, making an even application impossible.
4 Replace client's jewellery before the polish application to prevent affecting the finish.
5 Invite client to pay before polish application, so there is no risk of damaging the finished result.

Care Note

A film of oil will prevent the base coat 'keying' to the nail and will lead to the polish peeling in a very short time

Maintenance of Polish

The consistency of the polish to be used is a vital factor in successful polish application. If the polish is too thin it will run over the nail plate and into the cuticles. This results in a messy application and more coats will be needed to achieve an even coverage. It also produces a brittle finish which will split and chip easily.

If the polish is too thick it will be equally difficult to apply evenly. It will not dry evenly and will tend to 'peel' easily.

Special thinners should be used to correct the consistency of over-thick nail polish. If the polish is thinned with nail polish remover it will become totally unusable. This is due to additives in the nail polish remover that minimise the drying effect on the nail plate. If these are introduced into the polish they will adversely affect its consistency and drying properties.

Always use specially prepared nail enamel thinner from the same manufacturer as the polish being used. This ensures compatibility between the two. Add nail enamel thinner drop by drop. It is easy to over-thin the polish so always use the thinner sparingly. The polish should be shaken after the addition of every few drops, then tested for consistency. Continue adding drops of thinner and shaking until the correct mixture is achieved.

It is important that the polish is shaken by using a rapid rolling action between the palms of the hands. If the polish is shaken any other way, the sediment found at the bottom of most cream polishes will be disturbed, producing an uneven colour in the application. This is another reason for always storing the bottles of polish upright.

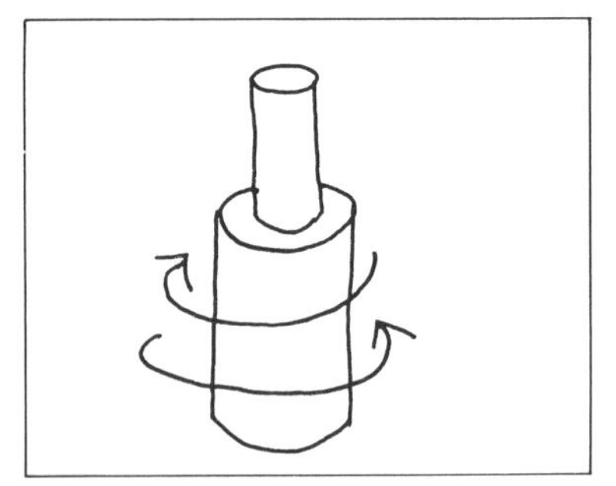
'Shake' only by rolling

Use ridge filler base coats, if necessary, before applying the polish. This helps to level the nail surface.

Work as quickly as possible when applying polish as this limits evaporation and maintains the life of the bottle of polish. Keep necks of bottles, stems and brushes clean, and close the bottles firmly.

Application of Base Coat

A base coat has the following functions:

1 It helps the polish to adhere to the nail.

2 It allows a more even polish application by minimising any ridges in the nail plate. Special 'ridge filling' base coats are available.
3 It prevents the pigments in the nail polish from staining the nail plate. This is a common problem encountered in manicure. If the client has used coloured nail polish without a base coat the nail plate becomes permanently discoloured. The staining cannot be removed and has to grow out. The client must be advised to use a base coat at all times, and if staining has occurred, a coloured polish will probably be required to hide the unsightly discolouration.
4 It prolongs the life of the polish application.
5 It helps to prevent the polish from chipping and peeling.

Method of Application

It is sensible to polish nails in an order that will help to prevent the manicurist from smearing completed nails when doing others. For this reason it is recommended that the right-handed manicurist works from left to right across the client's hands. Start on the client's left hand and work from the thumb across to the little finger. On the client's right hand, begin with the thumb, then pass to the little finger, and work back to the thumb. The left-handed manicurist should of course reverse this procedure.

There are a variety of methods of covering the nail plate with polish. Whatever method is adopted it must help to achieve the result of covering the nail plate evenly. The polish must not be allowed to run into the cuticles but produce a smooth line close to the cuticle.

Work in this order

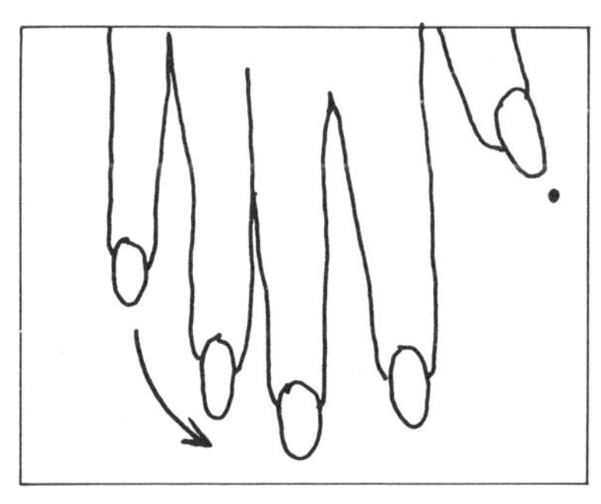

Client's right hand

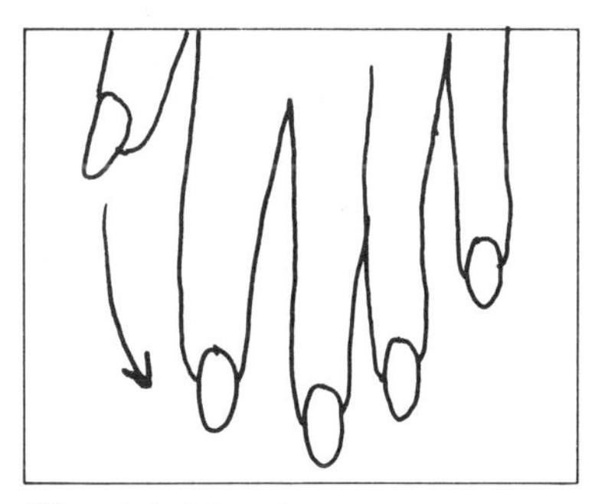

Client's left hand

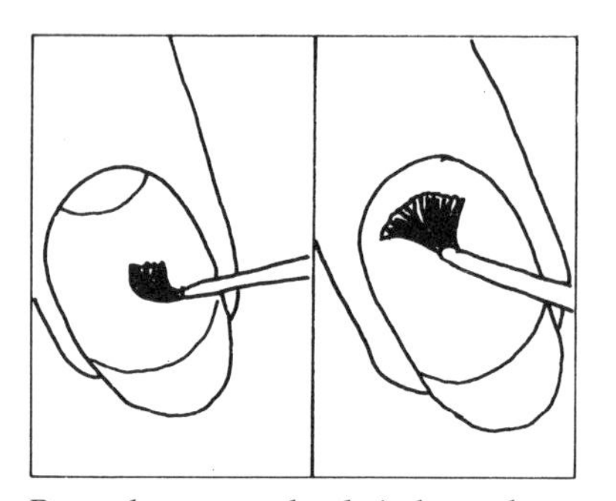

Press down to splay bristles and push in towards cuticle

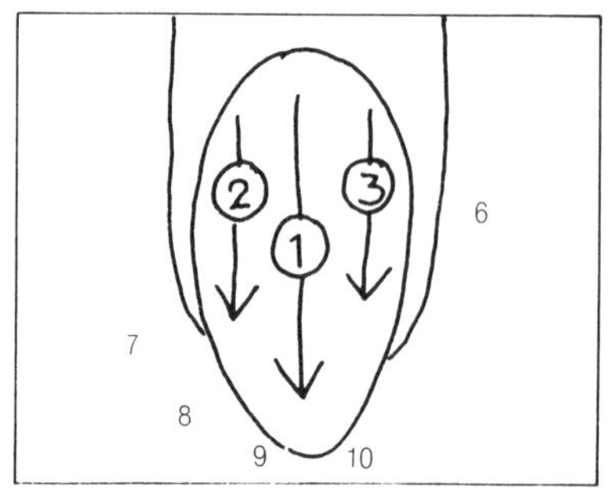

Polish application Method 1

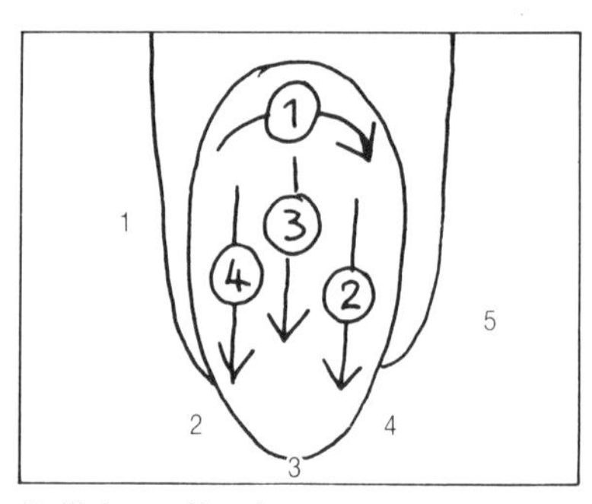

Polish application Method 2

Several methods are suggested in the diagrams above. It is advisable to experiment to find which is the easiest method to use and then keep to it. It will only occasionally be necessary to alter this method to accommodate particularly large or broad hands.

It is essential to develop a method of applying polish evenly with the minimum number of strokes. Too many strokes can produce an uneven result and make application more difficult because the polish begins drying – so speed is essential. This is one of the ways in which a competent professional manicurist can be judged. An examiner can easily identify a student using more than four strokes on each nail.

Use few strokes speedily

Apply the base coat evenly and accurately to all the nails, and under the edges for extra strength. Then proceed with the application of polish.

If the base coat is of the correct consistency there should be no need to wait until it dries before starting to apply the polish. (Polish of course, must be dry before proceeding further.) Do *not* apply any quick-drying sprays between coats because the oil in these will make further application impossible.

The choice of base coat is usually determined by the choice of polish. Some manufacturers' products are not compatible with others. Polish from one manufacturer may not go on easily over another manufacturer's base coat. Some base coats contain special additives designed to protect, strengthen or harden problem nails.

Application of Polish

Coloured polishes create a shiny, coloured coating on the nails. They are referred to by a variety of terms, such as

Check consistency of polish before beginning the manicure and adjust if necessary (p. 32)

varnishes, lacquers and enamels. They consist, broadly, of nitrocellulose which forms the film, resins to enhance adhesion and shine, plasticizers for flexibility and reduction of shrinkage, pigments for the variety of colours, and solvents for smooth mixture and application. Opalescent pigments such as guanine – derived from fish scales – are used in pearlised polish to produce iridescence.

Nail polishes must:

1. Be harmless to skin and nails
2. Be relatively easy to apply
3. Be reasonably stable in storage
4. Give a film of even consistency
5. Produce an even colour
6. Retain flexibility when dry to prevent chipping
7. Dry quickly and evenly without creating and trapping bubbles
8. Contain no toxic vapours. Nail polishes are sometimes 'sniffed' and could be unsafe
9. Give the client a wide colour choice.

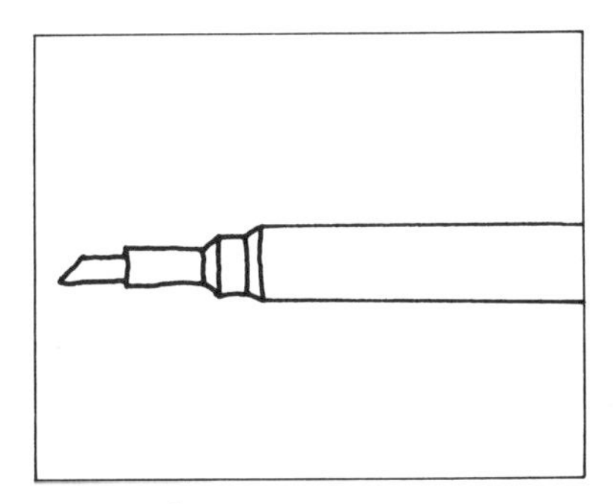

A varnish pen

Cosmetic chemists have paid a great deal of attention to nail polishes in recent years. Alternative products have been produced, such as a colour film that sticks to the nail like a transfer, and felt-tip pens that give a varnish-like colour. No doubt research will continue to produce even more products that will be long-lasting and easy to use.

Colour Choice

Choosing a colour remains an individual matter for the client. But clients often value advice on which colour might be most suitable. The nail varnish may be chosen to match an outfit, hair colour or lip shade. Obviously unsuitable shades should be avoided, such as dark cream colours on short stubby nails.

The colour used should always be recorded on the client's record card so that the manicurist knows which one to use on subsequent visits and can make sure it is ready in stock. At this stage the manicurist may ask if the client would like to buy a bottle of the chosen shade for use at home.

Opaline or Pearlised Polishes (also called frosted polishes)

These polishes are becoming extremely popular. The sheen produced is due to an additive derived from fish scales. These polishes also have the advantage of not requiring a top coat, which would only dull the finished effect. They do however require three coats, whereas cream enamel requires two coats and one top coat.

Method of Polish Application

The order of application should follow that described for a base coat (see p. 32), to ensure an even result and to prevent any smudging of the polish. The method best suited to the individual manicurist should be used. For strength and durability always use two coats of polish (except for pearlised polishes, see above). If it is felt that more than two coats are required then either:

1 The polish consistency is too thin
2 The colour selected was too light
3 The application of the polish was not even.

If too many coats are applied the polish will 'peel'. Allow the polish to dry between coats.

Half Moon Application

Some clients prefer the lunula or half moon left uncovered. Fortunately this is no longer fashionable as it is extremely difficult to achieve accurately. It remains an excellent exercise for students to develop their skills of coloured polish application.

Dark-coloured cream polish shows any flaws in the application so manicure students are advised always to practise with dark creams so that perfection is achieved. This also allows easier assessment of their application skills.

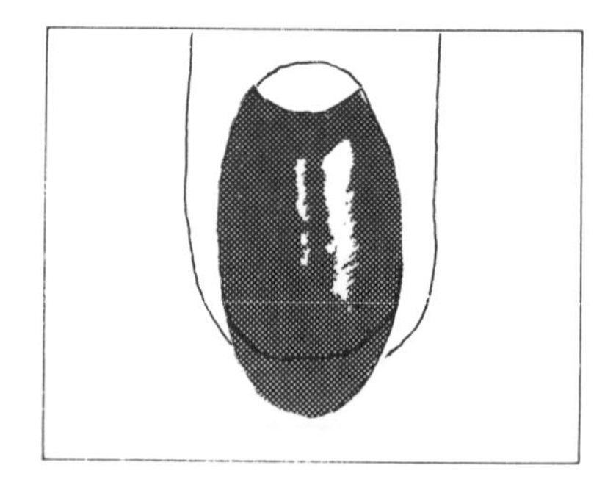

Half moon exposed

French Manicure

This is a system of using a natural pink colour polish on the main body of the nail plate and a much whiter polish on the free edge, to give a natural looking effect. The effect can often be easily achieved by just using a nail white pencil under the free edge – as long as the client has good circulation and the nail is a healthy pink colour.

Repairing a Smudge

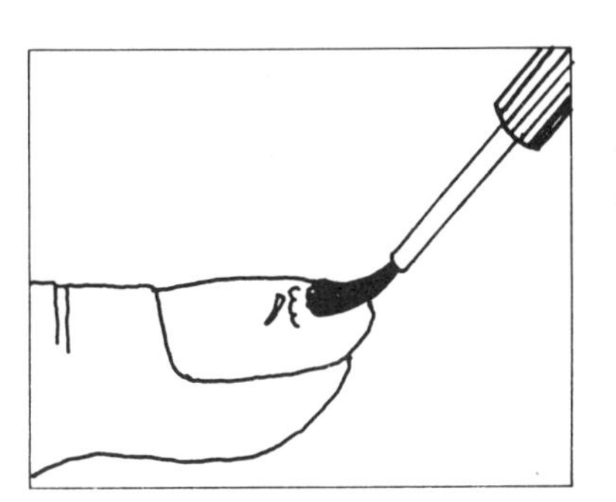

'Lift' polish over smudge

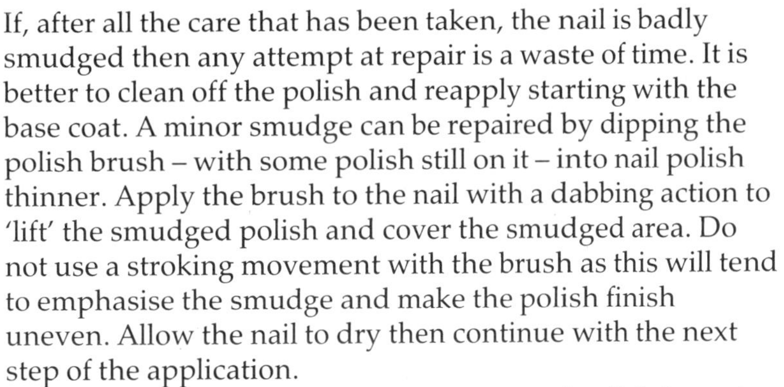

If, after all the care that has been taken, the nail is badly smudged then any attempt at repair is a waste of time. It is better to clean off the polish and reapply starting with the base coat. A minor smudge can be repaired by dipping the polish brush – with some polish still on it – into nail polish thinner. Apply the brush to the nail with a dabbing action to 'lift' the smudged polish and cover the smudged area. Do not use a stroking movement with the brush as this will tend to emphasise the smudge and make the polish finish uneven. Allow the nail to dry then continue with the next step of the application.

Some manicurists advocate removing nail polish from the hairline tip of the free edge to prevent chipping, but it is doubtful that this technique has much value. The polish will tend to 'wear' off the tip in a very short period naturally with use.

Removing Excess Polish

This is achieved by dipping an untipped orangewood stick into polish remover and applying it carefully to remove any excess polish from around the cuticle, deposited through faulty application. This is the only time in manicure where it is permissible to use an untipped orangewood stick. It should be applied with great care and very gently, to prevent spoiling the application and to avoid damaging the client's cuticle or matrix. Dip the stick into polish remover as often as necessary to remove excess polish without flooding the area being treated.

Special remover pens are available with disposable wedge shaped tips. These are expensive and are probably more useful as a retail item.

Top Coat

The top coat is a hard, clear coating that increases the shine of cream polishes. It prolongs the life of the polish and prevents scratching, flaking, chipping, and peeling. It acts as a sealer and can be used under the free edge as additional protection. It is applied with the same method and order of application as base coat and coloured polish.

Drying

It is essential that the polish is completely 'surface' dry before the client leaves because the new application can be very easily smudged and spoilt.

The nails should not be 'force' dried as this will cause bubbling in the application. Bubbles are brought to the surface where they burst. This sometimes happens when the manicurist is working by an open window or in a draught. It takes 20–30 minutes for newly applied polish to dry completely. Nails can be 'surface' dried in seconds by using a quick-drying spray.

Quick-drying Sprays

These are usually in aerosol form. They use the rapid evaporation of the aerosol propellant to dry freshly polished nails quickly. Sometimes drying is enhanced by depositing a light film of oil over the new polish surface to reduce tackiness and prevent smudging. Aerosols are usually sprayed downwards on to the nails at a distance of 12 inches (300 mm). Other products, such as paint-on liquids, are not as easy to use.

Unit 5

Massage

HAND MASSAGE

Massage is a method of stimulating and soothing the skin and muscles. The manicurist can use massage in the treatment of the hands, fingers, and arms. Massage can be applied by hand or machine, though hand massage offers a wider variety of movements.

Benefits of Hand Massage

The benefits of hand or manual massage of the hand and arm during a manicure treatment cannot be over emphasised.

- It is very relaxing for the client. This is often given as the main reason for having a manicure treatment.
- Joint mobility is very much improved. This applies particularly to the older client, and to those with early symptoms of arthritis. On no account should painful joints be manipulated as any problems could be aggravated.
- It aids blood supply to the hand and arm and improves lymphatic drainage. This in turn ensures the supply of the nail cell regenerative processes and maintains the strength of the nails.
- It stimulates and soothes the nerve endings in the skin.
- It smooths the skin surface and the rich emollient cream used improves both the feel and appearance of dry hands.
- Over a period of time muscle tone is improved.

There are several hand massage movements that may be applied to different parts of the body. Those of special use to manicurists include the following.

Effleurage. This smoothing, soothing, stroking action is performed with a firm but gentle movement of the hands and fingertips. It is used before and after most stimulating treatments. It improves the functions of skin and nerves, and relaxes tensed muscles.

Petrissage. This is a deeper, kneading movement used to break down fatty congestions, assist elimination of waste products, and the flow of nutrients to the tissues of the skin and muscles. There are several types of petrissage movements, e.g. kneading, pinching, pounding, friction, squeezing, and pressing. Of particular interest and help to the manicurist are kneading, light pinching, friction and pressing.

Tapotement. This stimulating movement consists of tapping or patting, hacking, and clapping. These are all quick beating movements applied with the fingertips. Tapotement may be used to stimulate nerves, restore muscle tone, and break down fatty deposits in the skin.

Vibration. This is a shaking movement applied with the hands or fingertips. Light vibrations are soothing and heavier ones are stimulating. These movements are imitated by vibratory machines. Stimulation of nerves, muscles, blood and nutrient supplies are achieved with these movements.

Joint manipulation. This technique aims to improve the mobility of joints in the fingers, hands and wrists. The joints must always be supported during manipulation. The client

Helpful Hint

The client may find it difficult to relax properly during the rotation movements of joint manipulation. The problem is often that the client is trying to 'help' by making the rotating movements. If this is the case, simply ask the client to look away and it will be found that the joint can be manipulated in the manner intended.

should be asked repeatedly if the movements are comfortable. Avoid this form of massage if the client has painful rheumatism or arthritis.

Friction. This brisk rubbing movement stimulates blood supply to the area treated.

There is no 'correct' sequence of hand massage movements. Many manicurists now use techniques associated with alternative therapy, such as reflexology, aromatherapy, acupressure, shiatsu, and Swedish massage.

Contra-indications to Massage

Massage movements should be checked frequently to ensure comfort.

Contra-indications are signs or symptoms which show that massage should not be carried out. Massage should not be given if there are signs of inflammation, breaks or cuts in the skin, spots, rashes, signs of disease, or if the client is undergoing medical treatment. To ignore contra-indications could result in harming the client, causing discomfort, and possible pain. Contra-indications should be noted at the initial consultation and fully discussed with the client.

It is important to use an emollient cream generously to help the fingers to glide easily over the massaged area with the minimum of friction.

Massage Routine

It is important to establish a routine that includes all the basic manipulations in a sensible order, for instance: effleurage – petrissage – tapotement – completing – effleurage.

The following routine has been developed with a commonsense approach. First the cream is spread thoroughly. The movements requiring the greatest fluidity (effleurage) are applied earlier in the sequence. Later movements allow the cream to be absorbed into the client's hands, while final movements 'use up' any remaining cream.

1 Take sufficient cream from the tube/pot and place in the palm of your hand. There should be no excess after massaging. The quantity needed is soon learnt through practice and experience.

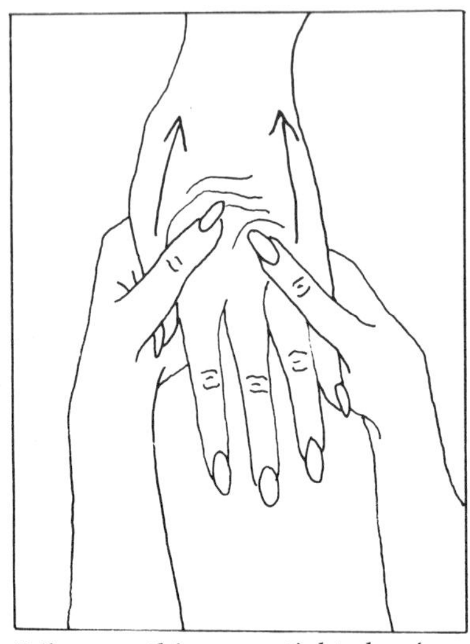

'Like smoothing on a tight glove'

2 Warm the cream by spreading it between your palms and fingers. Take the client's left hand and firmly spread the cream upwards and over the wrist. The action should be like trying to put on a very tight glove. (*Effleurage*)
3 Make circular movements with the ball of the thumb on the back of the client's hand. Work upwards towards the wrist then slide down to the base of the fingers. (*Petrissage*)
4 Turns the client's hand, palm side up, and repeat the action described in 3 but use a much firmer petrissage (kneading) movement.
5 Rotate the client's fingers, one at a time, five times clockwise and five times anticlockwise. Make sure that the client's hand is supported at the knuckle joint of the finger being worked on. (*Joint manipulation*)
6 Hold the client's fingers together and rotate the hand at the wrist. Move the hand in as large a circle as possible. Do this five times, in both clockwise and anticlockwise directions. (*Joint manipulation*)
7 The client's wrist should then be flexed by gently pushing the fingers back towards the arm. This should be done by the manicurist's fingers resting on the client's fingers. (*Joint manipulation*)

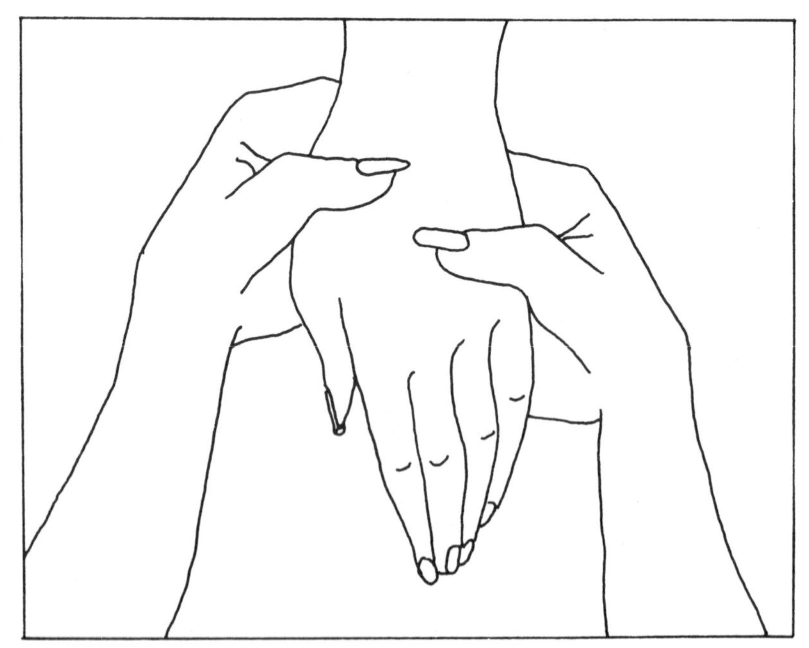

Effleurage—hand position

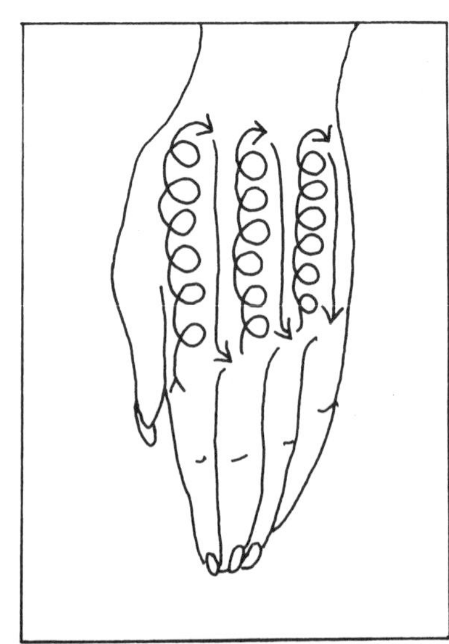

Effleurage—movement

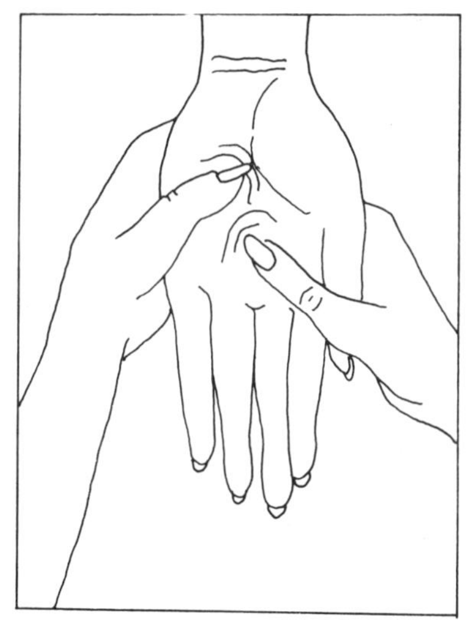

Petrissage—hand position

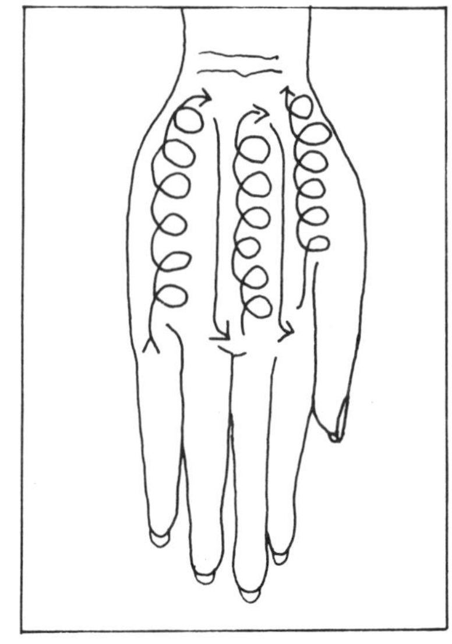

Petrissage—movement

Rotate fingers

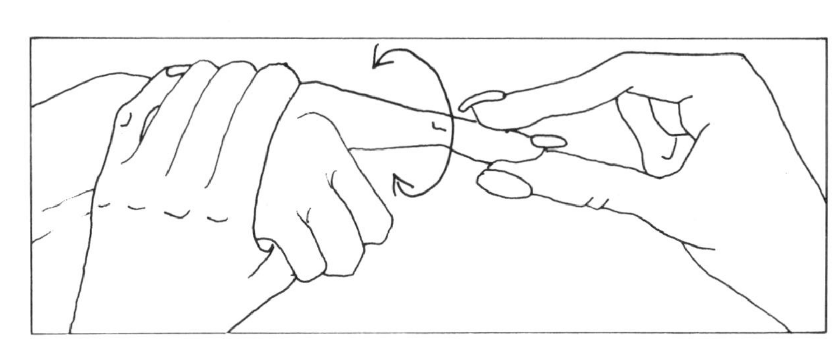

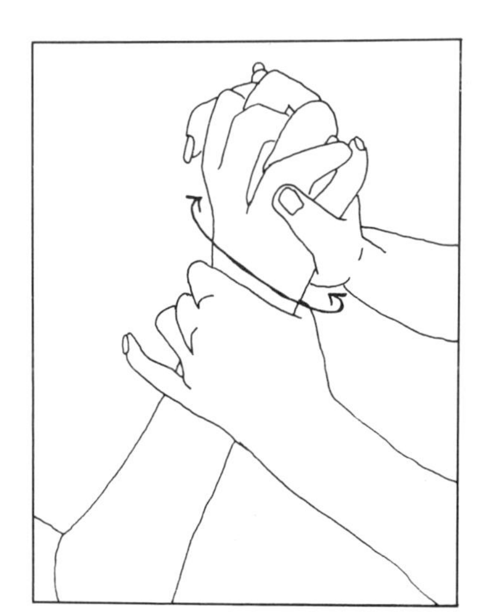

Rotate wrist

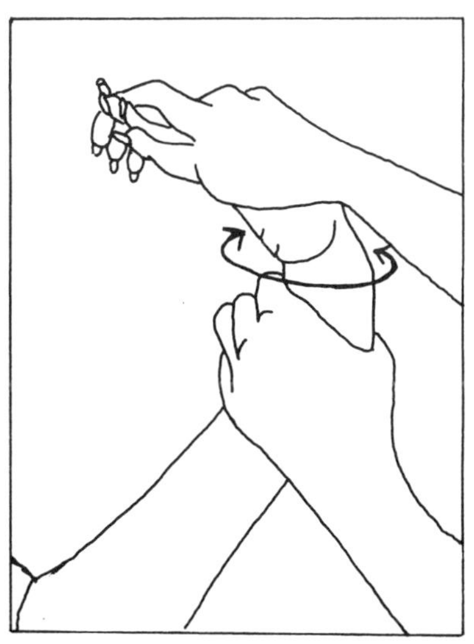

Rotate wrist with finger hold

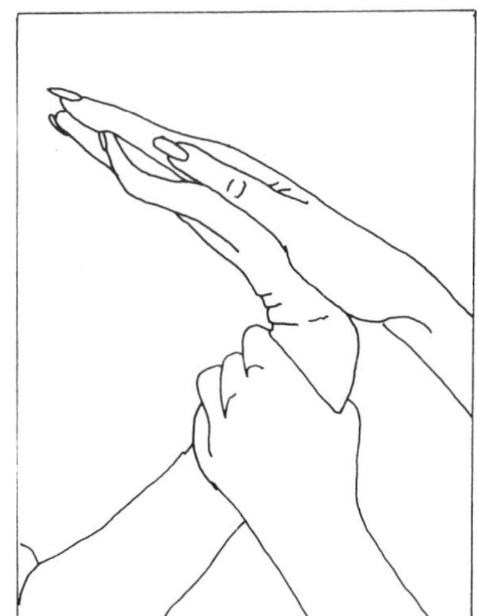

Flex wrist

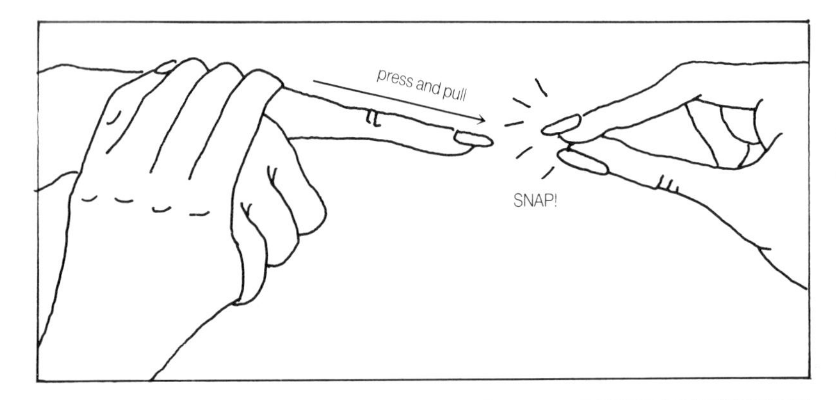

Finger massage—the 'snap' is the manicurist's finger and thumb coming together after the pressure on client's finger, NOT the client's joints!

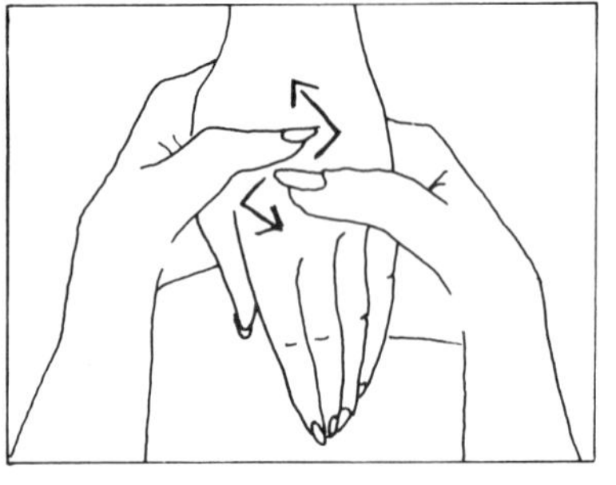

'Scissor' massage—thumbs change position with each other in zigzag motion, moving towards wrist

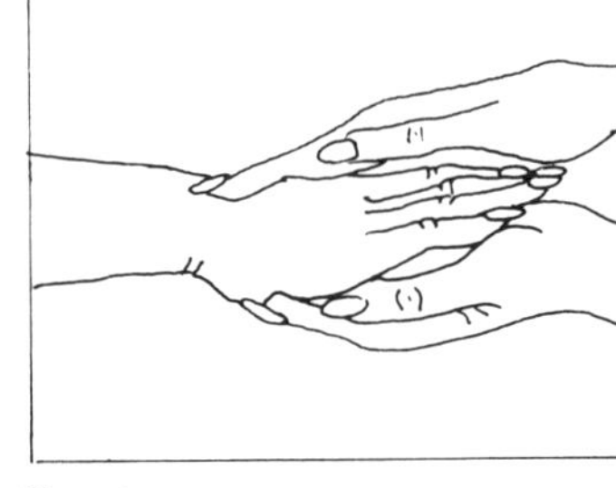

Tapotement

8 Firmly massage each finger with the ball of the thumb. Draw off the tips of the fingers. Repeat this five times for each finger. (*Effleurage*)
9 Massage the palms and the backs of the hands using a scissor-like movement of the thumb. (*Petrissage*)
10 Lightly and rapidly tap the client's hand between your own. This also helps use up any remaining cream.
11 Finish the massage with effleurage movements.

Note

When removing cream from pot or tube use a sterile spatula. Firmly replace lids or tops after use.

UNIT 6

Nail Repairs

There are several methods of repairing nails and the choice depends on the extent of the damage – a split on the free edge, nail wrapping, a break below the free edge or re-attaching a nail tip that has broken off completely.

Nail-mending materials usually contain a mixture of fast-setting adhesives, a fibrous reinforcing material and solvents. They are designed to allow a fast-forming film to adhere and support the nail. Several coats are applied in opposite directions to each other, allowing each to dry before further coats.

The film-forming material is usually nitrocellulose together with a resin, such as sulphonamide formaldehyde. Plasticizers are used for added flexibility, fibres for support, and the mixture includes quick-drying solvents such as ethyl acetate, butyl acetate and toluene, with dyes for nail colour.

Quick-setting repairers are available and are usually sold in an applicator to be used for emergency repair if a nail splits. These applicators can be convenient for clients to use at home if they are growing their nails longer, for a special occasion. The manicurist may find it useful to stock such kits for retail sales to clients.

There are however dangers associated with quick-setting repairers and both the client and the manicurist must be made aware of these.

- The repairer must not be used on an injured finger or inflamed nails in case of aggravation (see contra-indications in Unit 15).
- Because it is fast setting the repairer can easily stick skin together. There is no effective solvent and if skin is

accidentally bonded medical treatment will be needed.

- Contact with the eyes can cause permanent damage.
- The repairer must not be used where a tear or split exposes the nail bed as it will stick the nail plate to the nail bed.
- If used on the free edge repairer can give the nails an uneven thickness which cannot be removed. This will detract from the finished polish application.
- Repairer must not be used under the free edge as it can glue the hyponychium to the free edge of the nail plate.
- Some clients may be allergic to the product.
- Always examine the nails and areas to be repaired carefully and ensure that no contra-indications to treatment apply (see Unit 15).

Repairing a Split on the Free Edge of the Nail

This is probably the most common problem met by women with long nails. It always seems to happen just before an important event when the client wants her nails to be perfect.

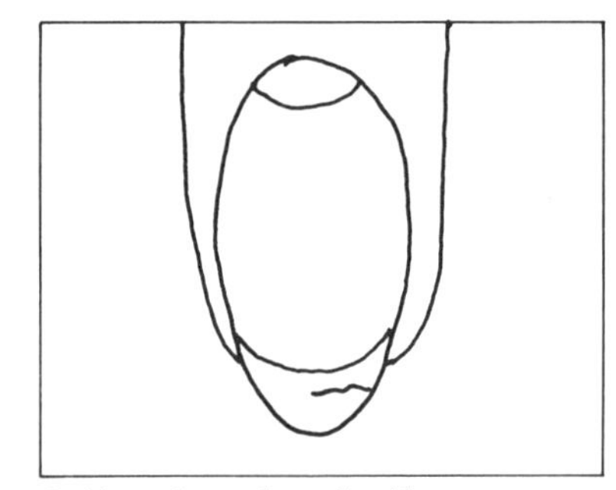

Split to free edge of nail

Method

1. Complete the manicure up to the base coat stage. Take particular care when filing not to make the split worse or to detach the free edge completely.
2. Clean the nail completely with nail varnish remover, to remove oil or cream. Make sure there is no oil or cream under the free edge. Any remaining oil or cream will prevent the repair patch from adhering properly.
3. Tear a piece of nail-mending tissue to the size required. The tissue must be torn and not cut, because a cut edge will show through the final polish. The mending tissue must be of high quality, fibrous, and absorbent. There are several manufacturers producing tissue specifically for this purpose. Linen or silk can also be used but the method is essentially the same.
4. Coat the nail and the underside of the free edge with nail-mending liquid. This resembles a clear base coat but has a thicker consistency.

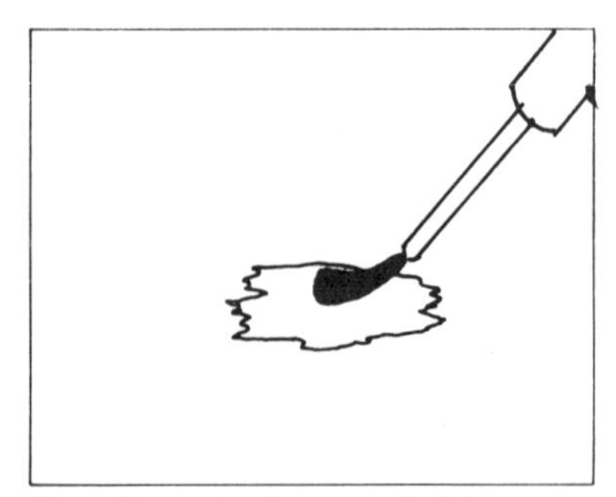

Tear tissue to shape and pick up with brush

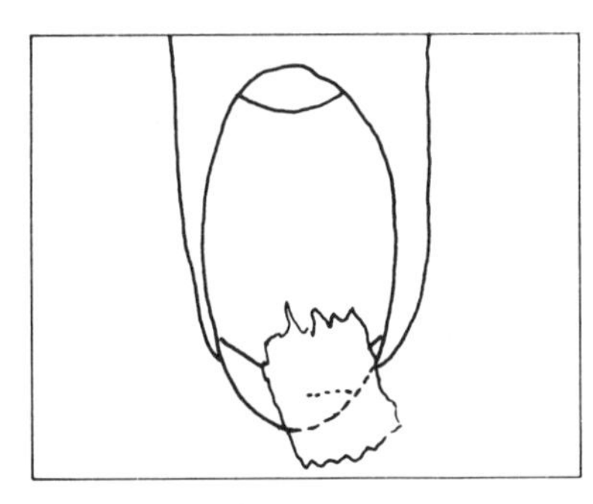

Place over split

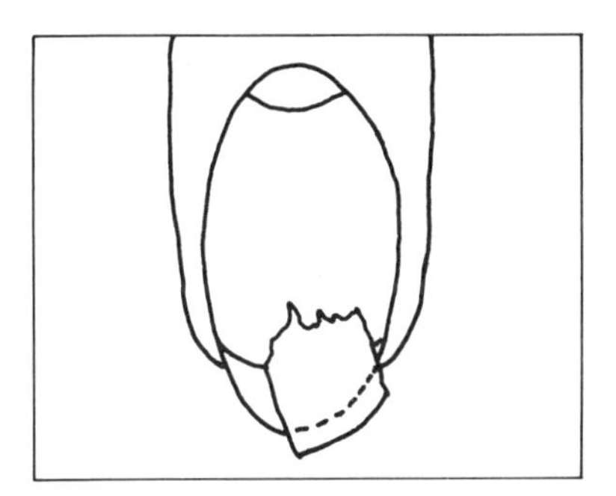

Trim excess

5 Coat both sides of the torn tissue with nail-mending liquid. To make this easier paint some nail-mending liquid on to the back of the hand, place the piece of tissue on to it, then paint the uppermost side of the tissue.
6 Remove the tissue patch from the hand with an orange stick and place it in position over the split nail. Dip the orange stick into remover beforehand to stop it sticking to the tissue. Roll the stick down the nail to remove any air bubbles.
7 Smooth the patch with a finger moistened with solvent. Always smooth away from the free edge to make sure the fibres in the tissue lie flat on the nail plate.
8 The overlapping tissue can then be trimmed if necessary, leaving enough to tuck firmly under the free edge, again using an orange stick dipped in solvent.
9 A buffer can be used to smooth out the patch further but this is not usually necessary unless silk or linen patches are being used for the repair. The patch must be totally dry before attempting this procedure or it will be re-lifted by the buffer.
10 Apply a coat of nail-mending liquid to the complete nail and under the free edge. Make sure this is dry before applying polish.

A good nail repair can last for several weeks and will survive careful polish removal. If the patch is still in good shape after removing the polish, leave the patch on and simply re-apply polish. Patches can be removed by generous application of nail polish remover.

Repairing a Nail that has Broken below the Free Edge

This process requires considerable skill and care. If done successfully it will save the client a lot of pain and discomfort because such breaks often tear across leaving sore, unprotected areas of flesh.

Method

1 Soak some cottonwool – which must be of high quality with short fibres – in nail-mending liquid. Tease out a

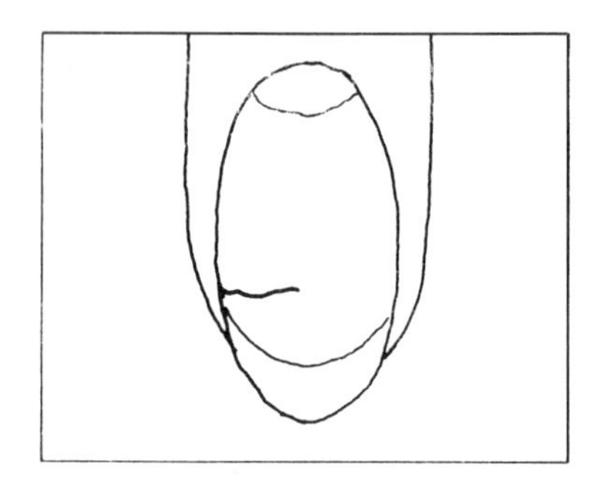

Split below free edge

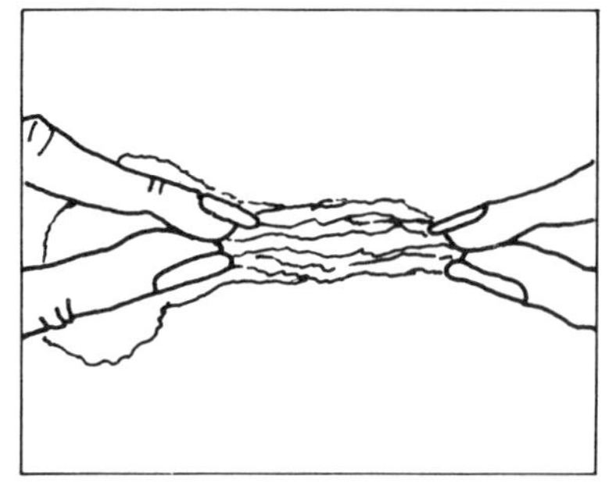

Tease out cottonwool into strands

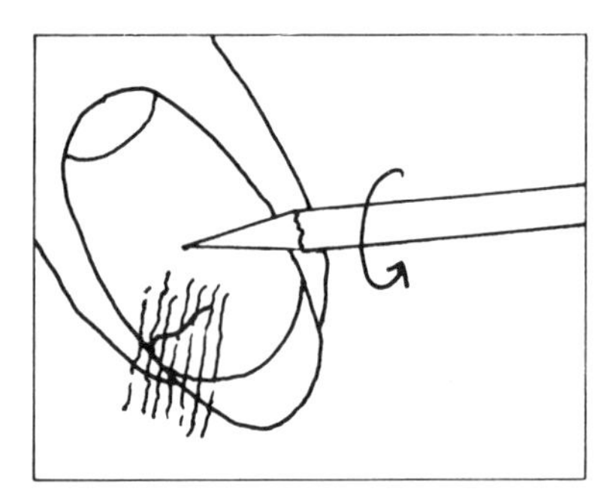

Lay across split

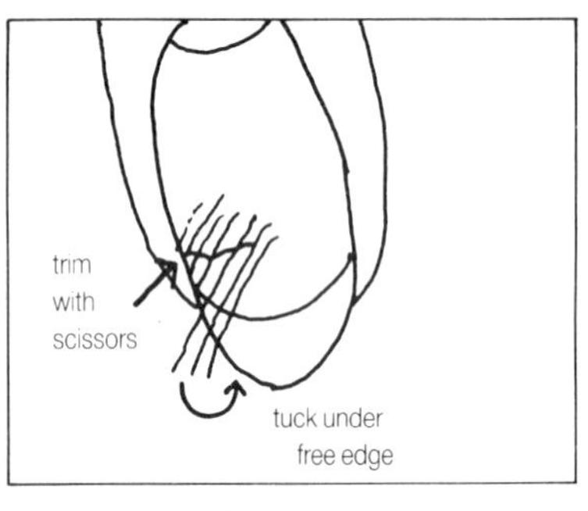

Trim and tuck under

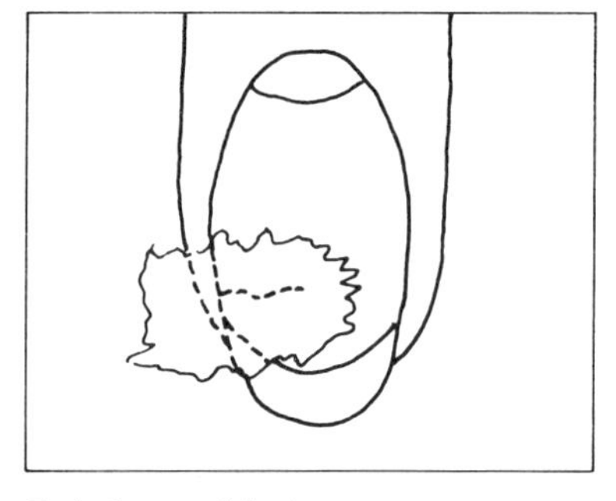

Reinforce with tissue

little cottonwool and lay it across the split diagonally, extending at both sides and beyond the free edge.

2 Lay the cottonwool flat on the nail by rolling with an orange stick previously dipped in solvent. It should be smoothed out carefully. Then trim the cottonwool with scissors and tuck it under the free edge.
3 When the cottonwool is dry apply a coat of nail-mending liquid to it.
4 Reinforce the break with a tissue patch using the same technique as in repairing a split on the free edge (see p. 46).

Nail Wrapping

Nail wrapping protects nails from splitting, peeling and breaking and protects the free edge from further damage. It is a very popular manicure service in the United States of America where women are, on the whole, more nail conscious.

Many different materials have been used, from torn

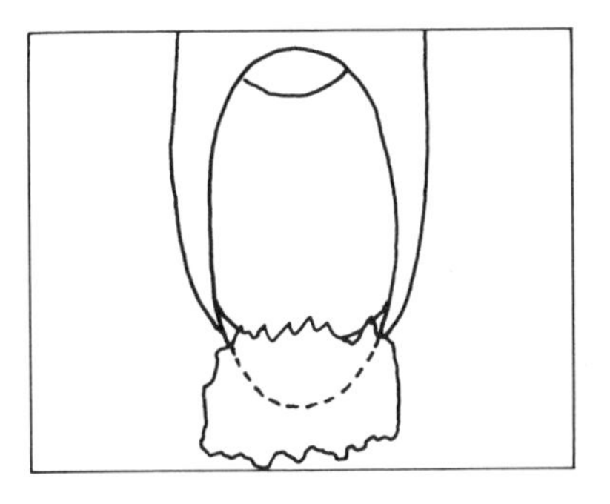

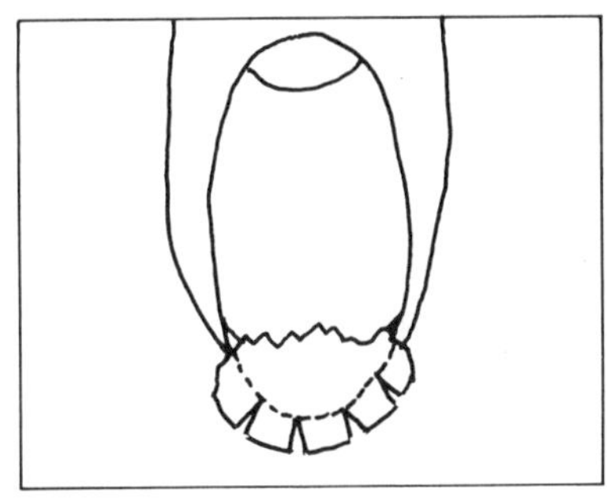

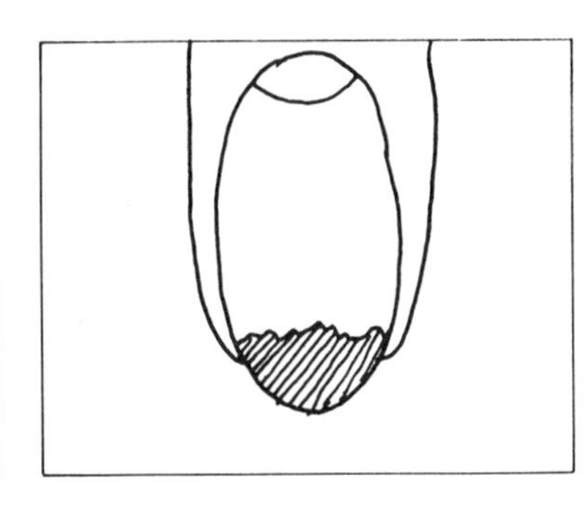

Nail wrapping

Trim edges and tuck under with orange stick

A wrapped nail

teabags to filter paper, but special tissues made for the task are usually the most successful. New products are constantly becoming available to protect nail tips. One recent development is a plastic sheet that adheres to the complete nail plate, after the polish application. Another is a liquid containing suspended fibres that is brushed on to the nail in a crisscross pattern to create a network of fibres.

The procedure for capping a fragile nail tip with tissue and nail-mending liquid is the same as for the split edge on the free edge. Particular care must be taken when tearing the tissue to size to ensure a perfect fit. The tissue must then be CUT to fit just under the free edge. It may be necessary to make additional cuts in towards the free edge so that the nail has a clean rounded edge on completion.

Re-attaching a Client's Nail Tip

Re-attaching a nail tip that has broken off completely is a complex and intricate task. It is very difficult to produce a smooth and even finish, and the results do not usually last for long. The client is probably better advised to have a more durable nail extension (see Units 8 and 9). However some clients prefer to have their own nails re-attached.

Alternatively the broken nail tip can be re-attached using nail extension techniques.

Method

1 Soak some cottonwool in nail-mending liquid. Lay it vertically along the nail plate with enough extending to

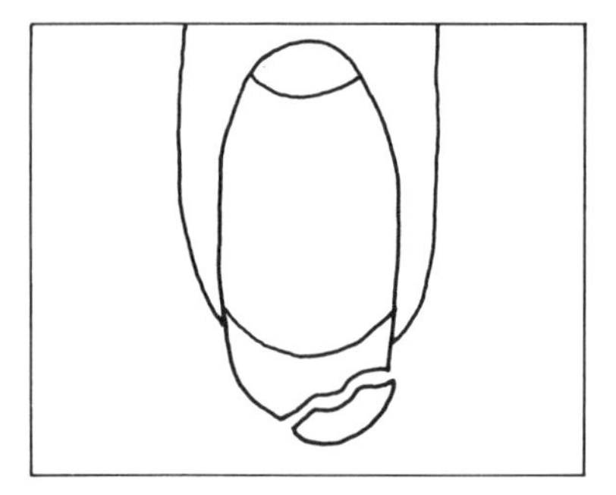
Detached nail tip

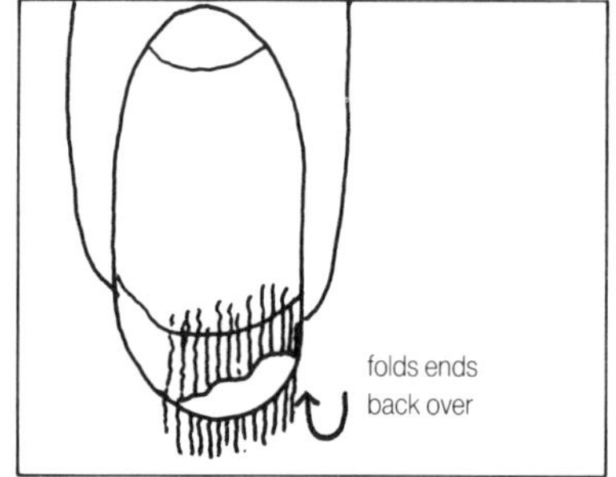

Place broken tip over cottonwool

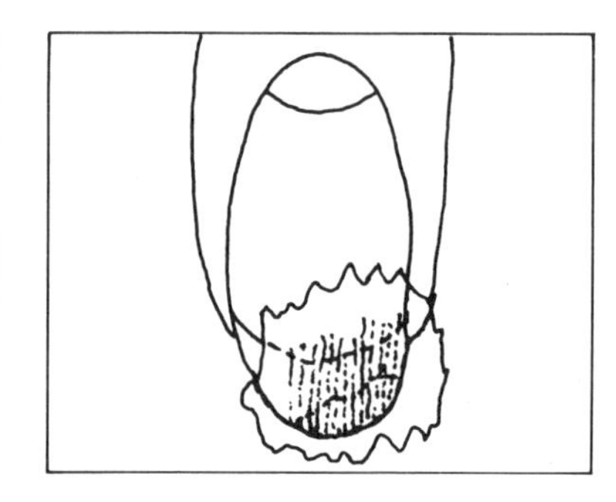
Reinforce with tissue

place the broken piece of nail in position on top of it. The overlapping ends of the cottonwool can then be folded back over the top of the nail.

2 Apply a coat of nail-mending liquid. When this is dry apply a reinforcing tissue patch horizontally across the break. Use the same technique as in repairing a break on the free edge.

Unit 7

Artificial Nails

Clients require artificial nails for a variety of reasons. The manicurist should recommend those most appropriate and likely to meet the clients' needs. (See client consultation p. 18.) There is such rapid development of manicuring systems, and so many new types of artificial nails, that some confusion exists.

The clients most likely to benefit from artificial nails are as follows:

- Nail biters who are unable to stop, need protection for the natural nail, help to prevent biting, and time to allow the natural nail to grow.
- Clients with broken nails that spoil the appearance of the hand.
- Clients with split, brittle, soft or weak thin nails.
- When the natural nails are mis-shapen.
- Clients who require fashion accessories such as gold nails.

Artificial nails can be roughly divided into three types:

- Temporary
- Semi-permanent
- Permanent.

It is not easy to make sharp divisions because different manufacturers market similar systems that vary according to the method of attachment or procedures followed.

Semi-permanent and permanent nails and extensions are dealt with in subsequent units.

Safety Tip

Many artificial nails, and the glues and gels used to attach them, are inflammable. They must be kept away from heat and flames. THE FUMES ARE TOXIC. Use in a ventilated room.

TEMPORARY NAILS

The traditional temporary nail, made of nylon or plastic, presses on to the natural nail. Clients are advised not to wear this type of nail for more than 48 hours. This avoids hindering natural nail growth.

The benefits of these temporary false nails are that they do not damage, soften, or weaken the natural nail plate. This is provided they are not worn for longer than 48 hours each time they are used.

Application Method

1. Examine the client's hands. Discuss the size and shape of nails required. Choose and match the false nails, and select the most appropriate.
2. Complete a basic manicure on the natural nails paying particular attention to the cuticles. Make sure there are no rough edges to stop the false nails fitting snugly.
3. Slightly roughen the surface of the natural nails with the fine side of an emery board.
4. If necessary file and shape the false nails at the cuticle end. Be careful to place them on double-sided adhesive tape in the correct order.
5. The bevel, or amount of curve, of the artificial nail should be matched to that of the client's natural nail. The bevel can be altered by placing the false nail in very hot water – not boiling – then altering it to match. Allow the false nail to cool and make sure that it is completely dry before adhering.
6. Roughen the inside of the false nail with an emery board to ensure better adhesion.
7. Apply a thin coating of the special slow-setting nail glue provided. Make sure the glue covers the area of the false nail that is to be in close contact with the natural nail – from the base to half way along the length.
8. Place the false nails, glue side uppermost, in the correct order. Leave until the glue becomes tacky. (See the manufacturer's instructions.)
9. Firmly press the false nails over the natural nail plate.

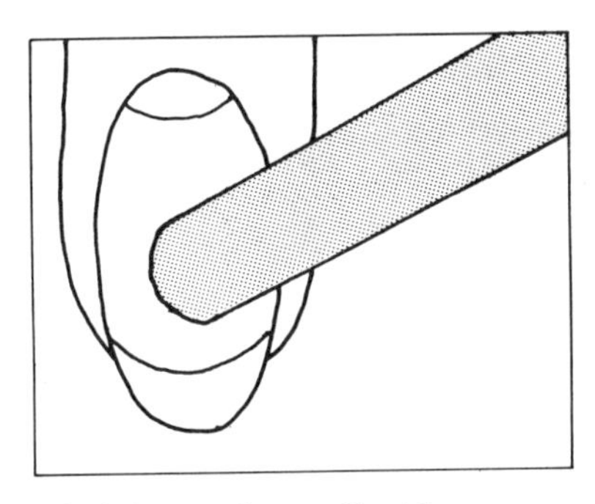

Slightly roughen nail with emery board

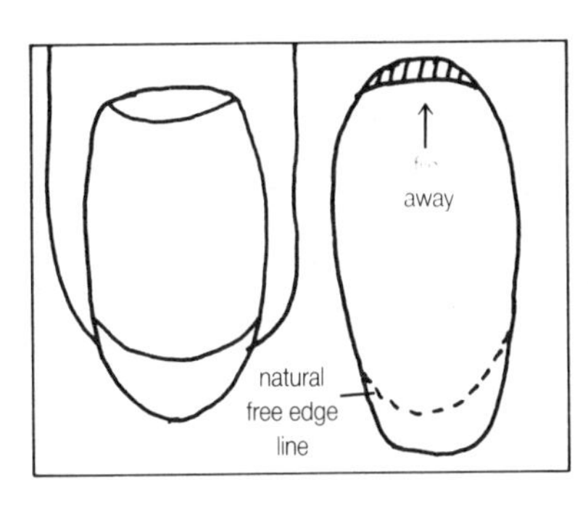

File to match natural nail shape at cuticle

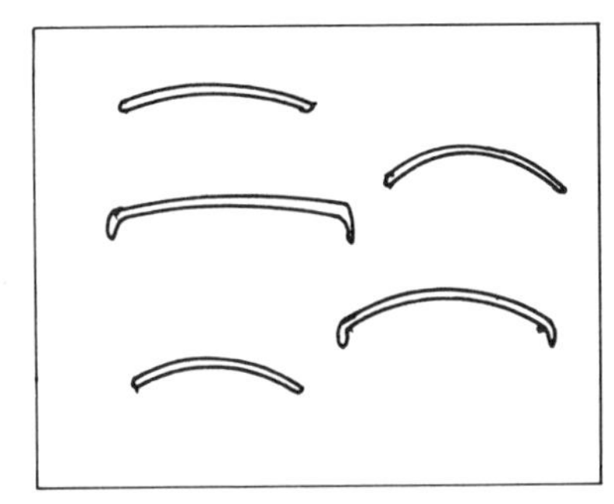

Sections through artificial nails bent to match curve of natural nails

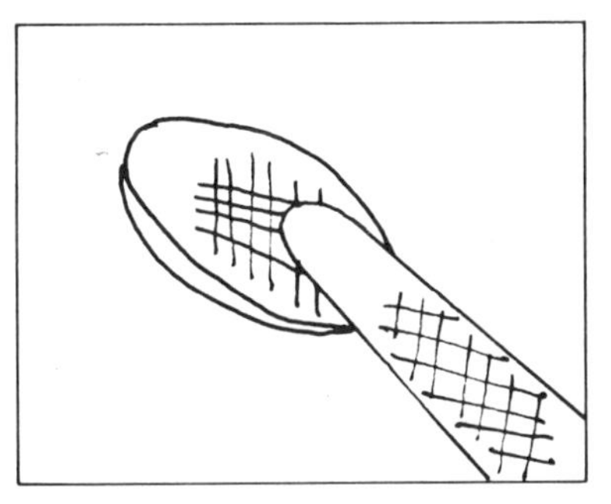

Roughen inside surface for better adhesion

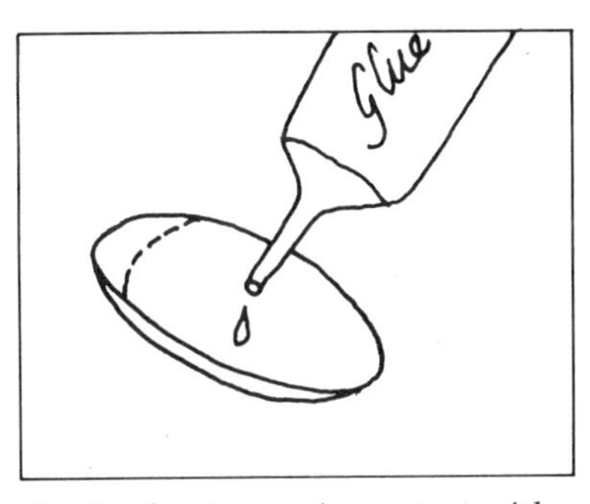

Apply glue to area in contact with nail

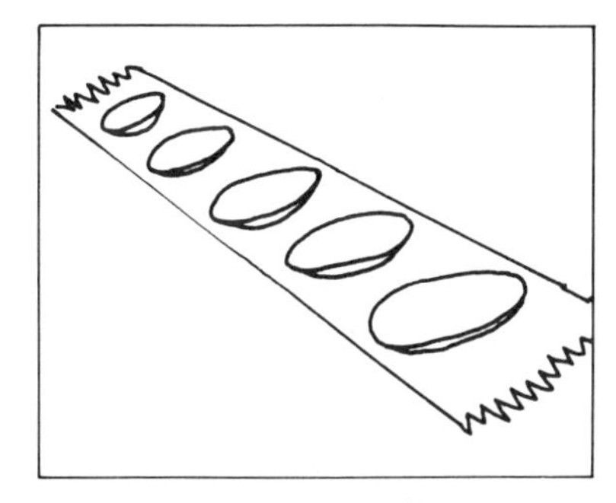

Place on double-sided adhesive tape in correct order

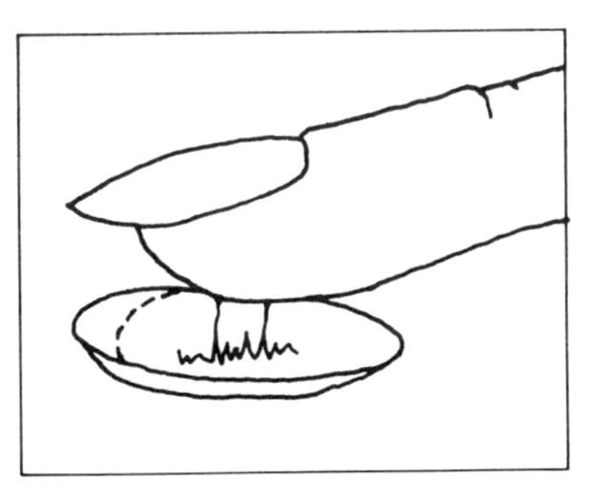

Leave until tacky

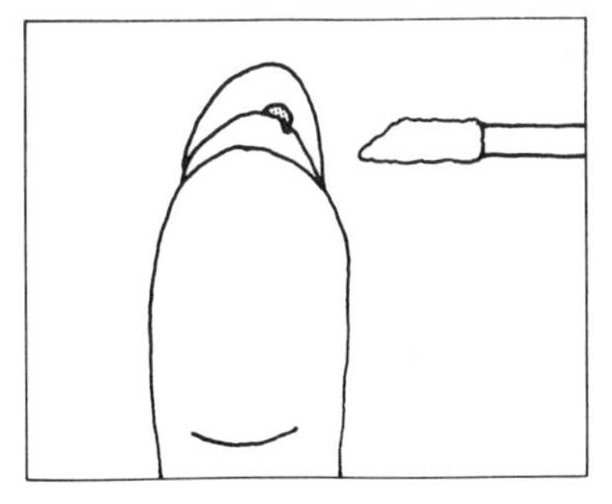

Remove excess glue with solvent

Advise the client to treat them very gently for another 30 minutes until the glue has completely set.

10 Remove any excess glue with solvent. Apply base coat, polish and top coat in the usual way.

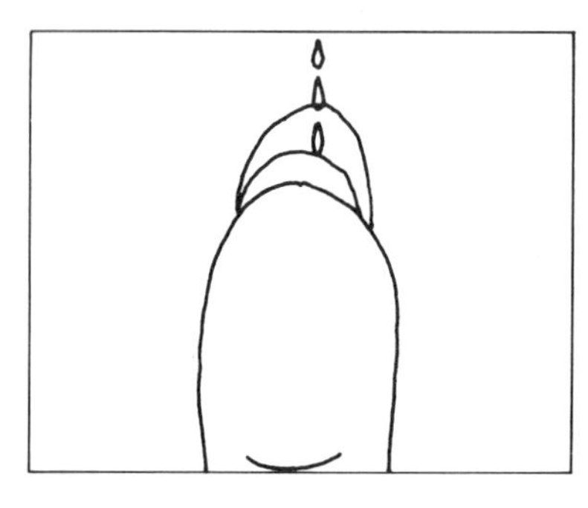

Apply solvent between false and natural nail

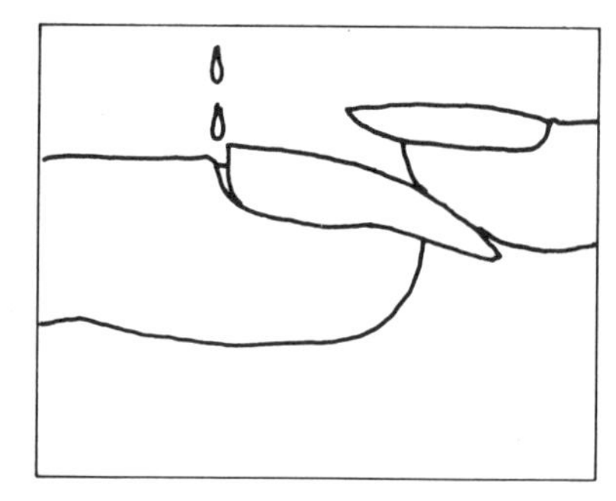

Rock gently—add more solvent

Removal Method

To remove this type of false nail, use either an 'acetone-free' polish remover or a special glue solvent provided by the manufacturer. The remover is applied, drop by drop, between the false and the natural nail. The false nail is gently rocked and more polish remover added. Eventually the false nail will loosen and come away. The back of the false nails and the surface of the natural nails can then be cleaned to remove the glue. The false nails can be used again if required.

Care Note

If an acetone-based polish remover is used it will dissolve the plastic or nylon false nail making it unfit for further use.

Instant Press-on Moulded Nails

An updated version of the press-on moulded, plastic nail, is now available. It is attached to the natural nail with double-sided 'tabs'. These nails have the great advantage of being quicker to apply – no time is spent waiting for glue to set. There is also less possibility of the false nails being dislodged or moved.

Application Method

1 Repeat steps 1–3 for the application of traditional moulded false nails.
2 Cut the tab to fit the natural nail. Then carefully peel it from its backing, holding one corner with pointed

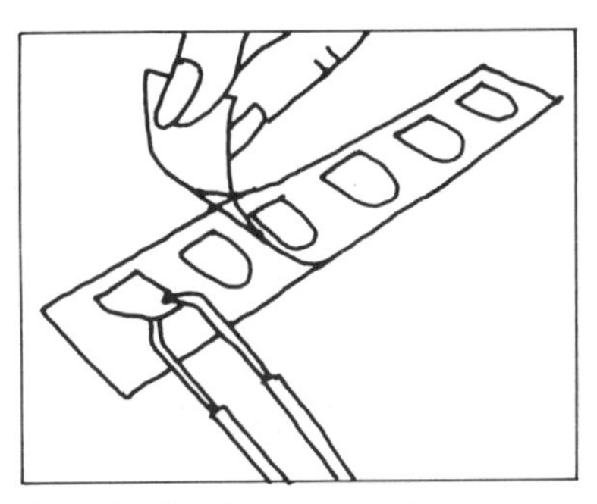
Double-sided adhesive tabs — use tweezers

tweezers. Avoid touching either of the sticky sides as this reduces its strength of adhesion.

3 Apply the rounded end as closely as possible to the cuticle. Press down firmly. Make sure that pressure is applied equally to all areas of the nail.
4 Apply double-sided tabs to each of the other nails.
5 Peel the second backing from the tab on the first finger – using tweezers. Again avoid touching the sticky film.
6 Select the false nail previously prepared for the first finger.
7 Place the false nail on to the natural nail and sticky tab, matching the cuticle, and press firmly.
8 Apply the false nails to each finger as required.

If the false nail has not lined up properly with the cuticle simply remove the false nail, then clean both the natural and false nails using an acetone-free polish remover, or recommended glue solvent. Apply another double-sided tab and re-attach the false nail.

Though not suitable for housework tasks such as washing up or other vigorous activities, these false nails are ideal for temporarily, and conveniently, giving the appearance of long, elegant nails.

Clients are advised to buy extra supplies of tabs to carry out repairs as required.

With care these nails can be used many times. A client who has become confident in the use of them will want to apply them at home themselves. They are available in a wide variety of colours and do not require polish application. Retail packs are also available. And packs of extra double-sided tabs will add to the manicurist's sales.

As with traditional temporary nails, clients are advised to remove these press-on nails after 48 hours. This prevents softening and damaging the natural nail plate.

Removal Method

This is a simple process. Soak the nails in warm soapy water, then gently rock the false nail until it becomes free. Alternatively acetone-free polish remover or the recommended solvent can be used. Remove surplus glue by rubbing with the fingertips. Finish by cleaning the natural and false nails with acetone-free remover or glue solvent.

Unit 8

Semi-permanent False Nails and Extensions

The term 'nail extension' can be applied to any system of lengthening the natural nail. But it usually refers to semi-permanent or permanent methods.

Semi-permanent false nails are expected to stay in place for up to two weeks, by which time the natural nail will have grown, leaving an unsightly gap at the cuticle end.

The method of choosing, matching and preparing the nails is as described for temporary nails in Unit 7 (p. 51).

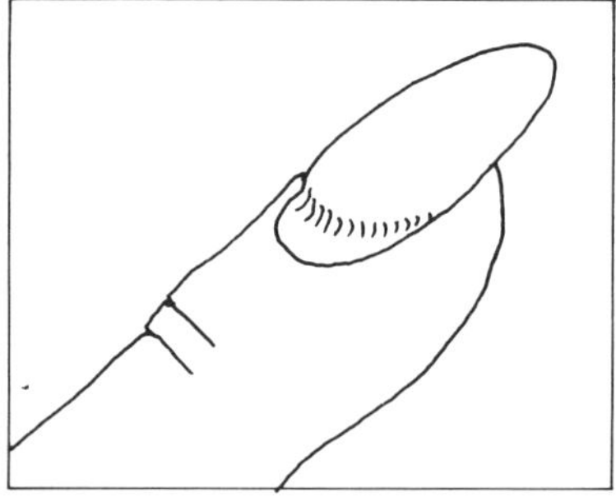

After 2–3 weeks' growth

Warning

The glues used in this process usually contain cyanoacrylate which creates an instant and very strong bond between surfaces. It is essential that the client's skin and cuticle are kept clear of the glue. Care is also needed to prevent the manicurist from becoming stuck – either to him/herself or to the client. If such an accident occurs see guidelines below.

Care Note

Only use glue that is specially prepared for nails. Other preparations can generate very high temperatures when drying and this can damage the nail and be extremely painful and uncomfortable for the client.

Accidental Adhesion of Skin Surfaces

Most glues for semi-permanent nail extensions contain cyanoacrylate – a strong, fast-setting adhesive. Skin can become bonded in seconds. Non-surgical first aid is considered the best treatment. The following guidelines will help if an accident occurs.

> **Warning**
>
> Seek medical attention immediately for accidents involving the eyes or eyelids.

Skin Bonding to Skin

The areas affected should be soaked in warm soapy water. Gently turn the surfaces apart with a spatula. The adhesive can then be removed with soap and water. DO NOT force the surfaces apart.

Eyelid to Eyelid, Eyelid to Eyeball

Medical attention must be sought immediately if cyanoacrylate glue gets into or near the eyes. As a first aid measure, bathe the area with warm water and cover.

Lips

If lips stick together, keep them wet with both saliva and warm water. This will help to part them in 12–48 hours. The lips MUST NOT be prised apart forceably.

The adhesive solidifies very quickly and is difficult to swallow – it sticks to the mouth first. Once pieces of the adhesive are freed care must be taken not to swallow any.

Burns

Cyanoacrylate gets hotter as it solidifies and can cause burns. These can be treated after the glue lump is released from the surface.

Surgery

Surgery is not thought necessary for separating bonded skin surfaces.

Application Method

1. The manicure should be completed up to the stage of polish application. Make sure the cuticle is completely loosened from the nail plate and pushed back. This allows the false nail to match its base and stops it looking obviously false.

2 Lightly buff the surface of the natural nails and the underside of the false nails. Some manufacturers suggest that the dust should not be removed as it helps the bonding. Others insist that both surfaces are scrupulously clean.
3 Work from left to right if right-handed, or from right to left if left-handed.
4 Take the first nail to be applied. Check it matches the nail shape and cuticle exactly. If the false nail does not exactly match the client's nail, it can be adjusted (see p. 64).
5 Apply a fine layer of glue to the nail plate. Work very quickly but make sure the nail is completely covered. Avoid glue becoming attached to the cuticles.

DANGER—toxic fumes!

Note

Most glues of this type are supplied with fine applicator nozzles. A very fine sable brush is more accurate and even in use. If the bristles are not sable the glue will destroy them, causing uneven glue patches on the nail plate. The metal securing the bristles should be seamless to give a firm hold. Sable brushes are expensive, but they give a high-quality and durable finish. They are available in all shapes and sizes. Soak sable brushes in solvent after use.

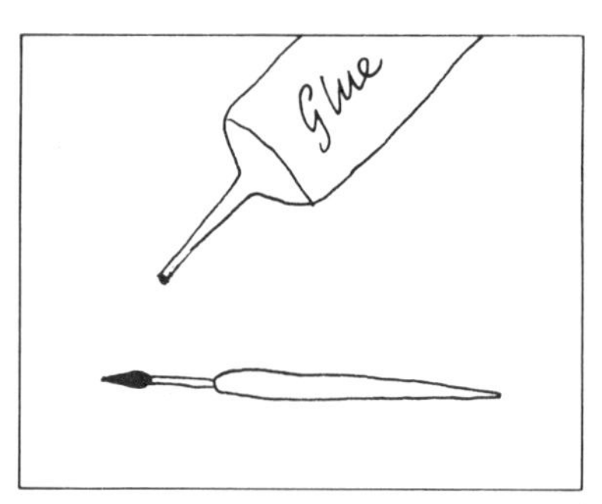

Fine nozzle applicator and fine sable brush

6 Apply the false nail to the nail plate using tweezers, positioning it as close to the cuticle as possible. Press the false nail firmly into position. It is a good idea to place a clear plastic film between the nail and the manicurist's fingers – glue will not bond to the plastic. It does become

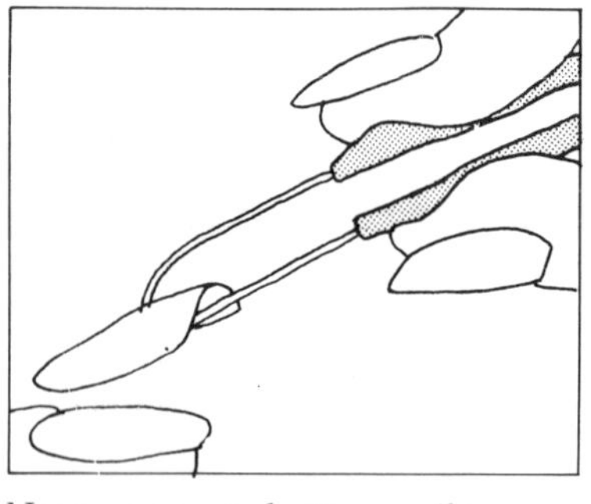

Use tweezers to butt up nail to cuticle

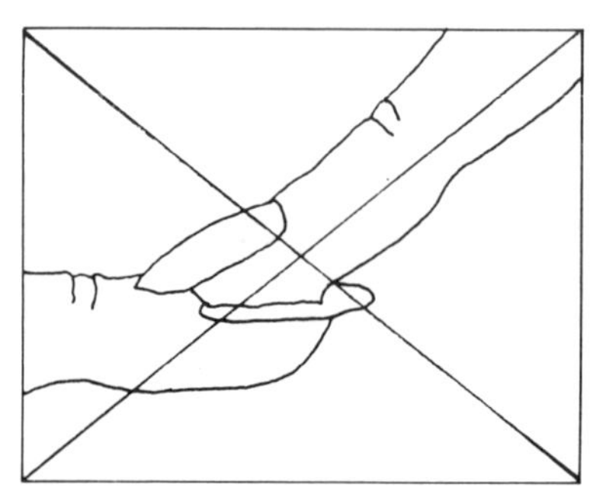

DON'T stick fingers together!

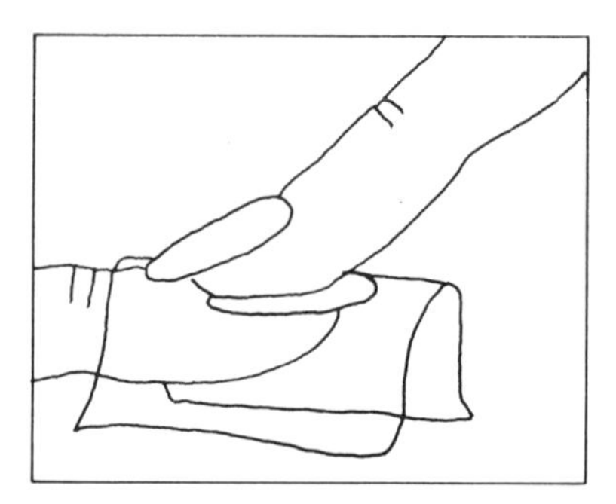

Use plastic film

> **Note**
>
> If the client and manicurist should become stuck together with cyanoacrylate glue, gently separate fingers with a wooden spatula. DO NOT FORCE APART.

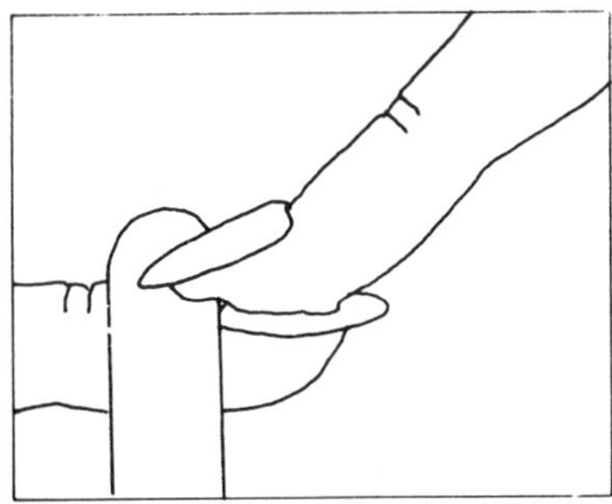

Gently separate with spatula

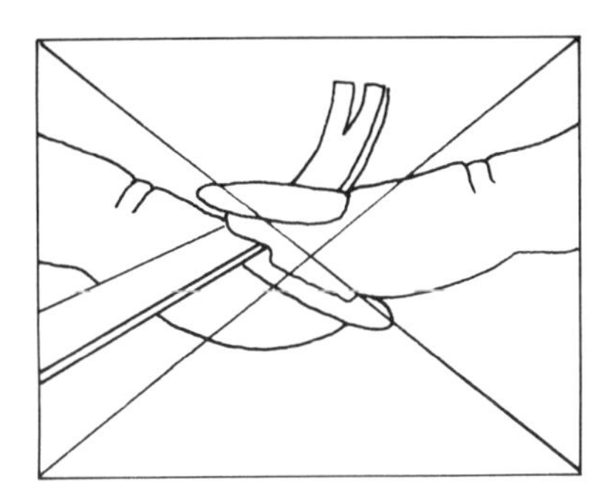

DO NOT force apart

> **Care Note**
>
> A client's confidence is soon shaken if expensive false nails come off in a few days. Accurate and careful application helps to avoid this.

embarrassing if the manicurist becomes firmly stuck to the client.

7 Apply the other false nails in the laid-out sequence.
8 Apply base coat, nail polish, and top coat in the usual way.
9 Make sure the client has information on caring for the new nails.
10 It is important to stress that the artificial nails should not be worn for longer than three weeks because the bases of the natural nails will be noticeable by then.
11 Offer another appointment for three weeks time.
12 In some cases a nail does not stick firmly, or it comes off. This is usually due to incorrect preparation. Warn the client of this very small possibility. Explain that should this occur they should return to the salon as soon as possible to have the nail re-applied – free of charge.

Removal Method

1. Cut away as much of the free edge as possible. Take care not to cut the natural nail below. It is important to keep the natural nail as long as possible to give a firmer bond to the refitted false nails.
2. Using a very fine eye dropper or sable brush apply a few drops of the recommended nail glue solvent between the artificial and natural nail. Apply to the base of the nail and gently rock the false nail until it is easily removed. Take care not to cause discomfort.
3. Use cottonwool soaked in glue solvent to clean any glue from the nail plate.
4. Continue until all the nails have been removed. Wash the hands and nails thoroughly with liquid soap and warm water. Use a soft nail brush under the free edge.
5. Restore the natural shine of the nails using a buffer and buffing paste.

Care and Maintenance

Clients should be given the following advice on caring for semi-permanent false nails.

- Treat the nails as if they were natural – carefully.

- Wear protective rubber gloves for all household tasks, for instance when preparing vegetables or using strong detergents.
- Do not be tempted to use the nails for peeling labels.
- Use finger pads when picking up objects where possible. Use the whole hand to lift heavier items.
- Regularly use a nail brush – gently – to remove dirt from under the free nail edges.
- Do not pick at, or lift, the edges of false nails.
- Frequent or repeated polish applications should be avoided. (Polish does not chip as easily as it does on natural nails.)
- As the nails grow away from the cuticle, after 2–3 weeks, a portion of the natural nail becomes visible. It is essential, at this stage, to decide whether to replace the nails completely, remove them completely, or have the gap filled with a gel or acrylic system.
  DO NOT be tempted to extend the period of wear. Water and bacteria could penetrate the gap and fungal growth occur between the natural and artificial nail.
- Attempting to remove the nails by picking, pulling or lifting them will almost certainly damage the natural nail. This damage could take many months to grow out.
- Always return to the salon to have artificial nails removed with special solvents.

Care Note

A darker-coloured polish can be applied simply by covering the existing one. If it is necessary to remove polish use only acetone-free remover.

UNIT 9

Permanent Nail Tip Extensions

This technique uses specially prepared, moulded nail tips. They are available in a wide selection of shapes and sizes to match any natural nail. The simplest and easiest method of applying them is by *gluing*.

They consist of plastic with a thinner inside 'ledge'. This allows them to fit over the natural nail tip. There are many types available. Expect to find that those with a 'thinner ledge' will be more expensive. These will not require as much buffing down at the joint.

Systematic storage of plastic tips is essential if fast and accurate selection of appropriate shapes is to be made. Different shapes are required for different fingers and no two hands are exactly alike. If the exact size and shape required is not available go to the next larger size and trim to fit with scissors.

A small sliding drawer unit provides a practical storage system. This can be of the type available in hardware

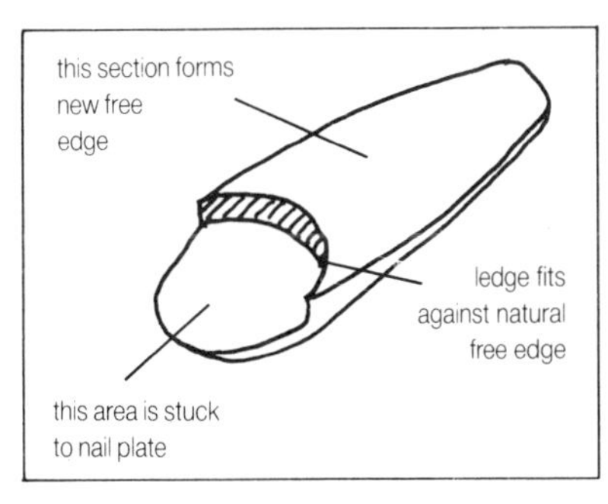

Exaggerated scale shows construction of nail tip from below

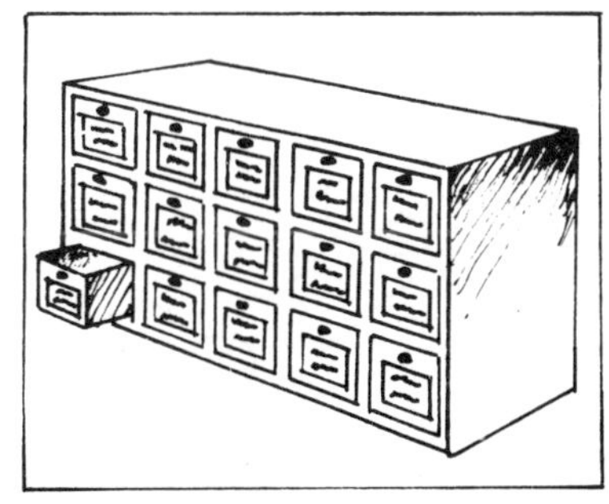

A storage cabinet

departments for storing small screws and suchlike. The drawers are easily labelled with the shape and size of nail tip enclosed. This is also useful for stock control since the stock level can be efficiently assessed and kept up to date.

> **Note**
>
> The following technique is *not* suitable for clients with very small nails, bitten nails, or nails without some free edge. The area of attachment will be too small to ensure good adhesion.

Application Method

1. Complete the manicure to the stage of polish application.
2. Select plastic nail tips in appropriate shapes and sizes.
3. Set out the tips in correct order on double-sided adhesive tape.
4. File the free edge to the same shape as the ledge in the plastic nail tip. If using nail tips with thicker ledges file the overlap to reduce the thickness.
5. Roughen the natural nail surface where the ledge is to be attached.
6. Apply glue to the part of the natural nail where the tip is to be applied. Carefully place the tip so that the ledge matches the free edge of the natural nail. Point the finger down so that any excess glue runs towards the free edge. If possible apply another drop between the underside of the tip and the natural free edge for greater strength.

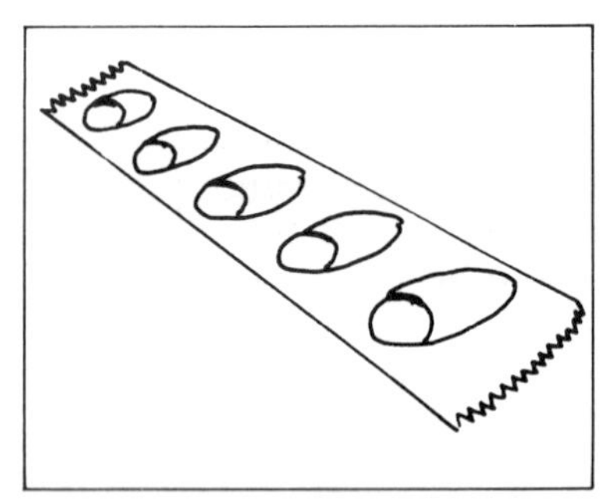

Lay out in correct order

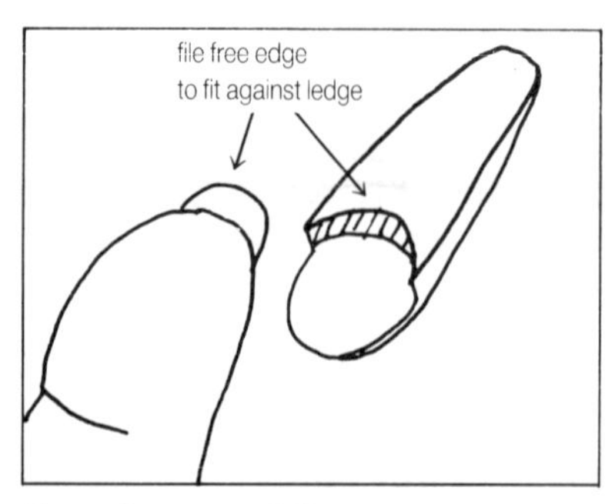

Ensuring a good fit

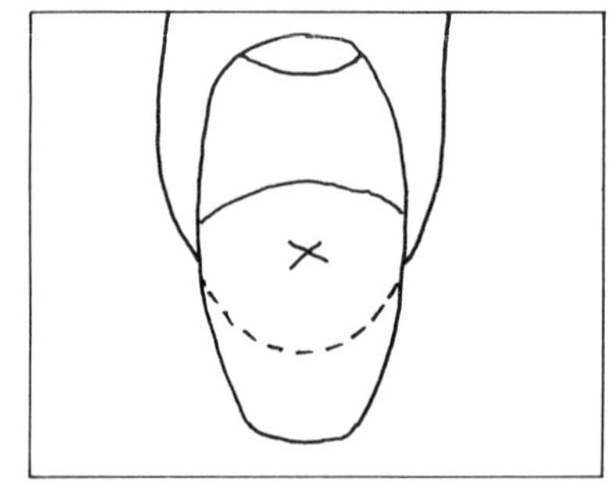

Glue here

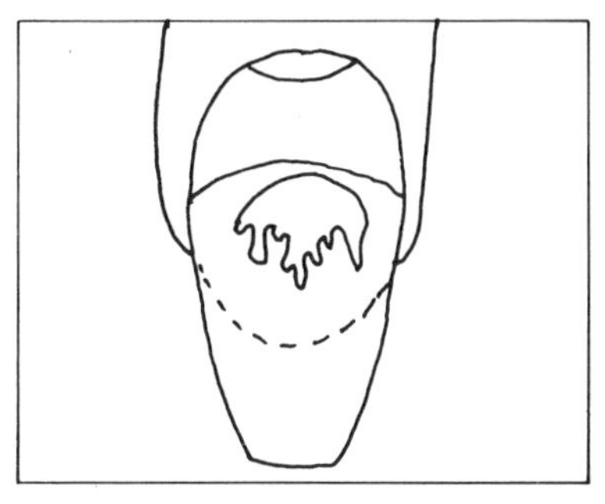

Hold finger down to let excess glue run to free edge

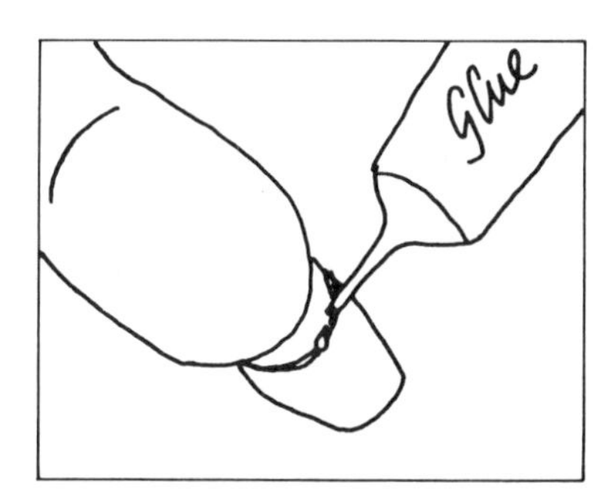

Glue free edge/platform joint for extra strength

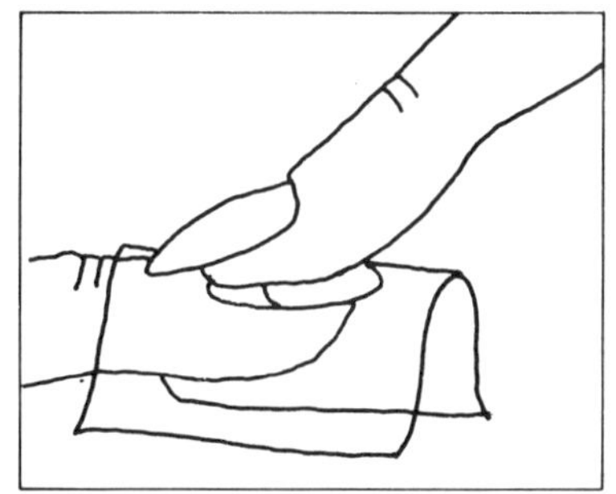

Press to spread the glue using plastic film to stop manicurist sticking to the client

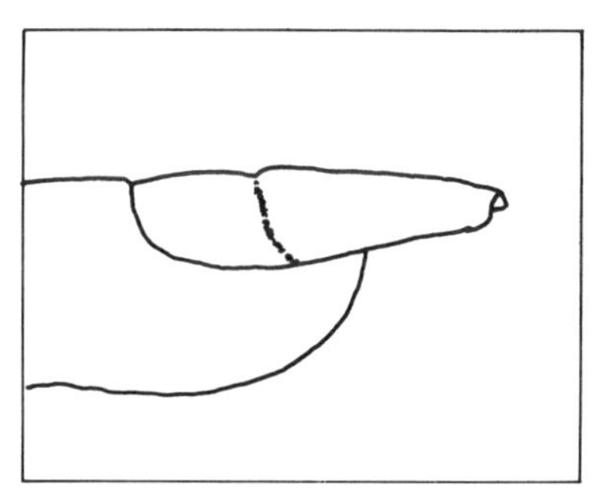

Keep tip parallel

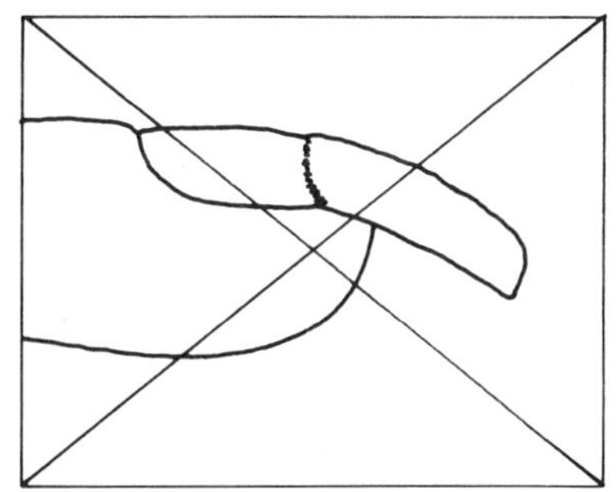

NOT like this

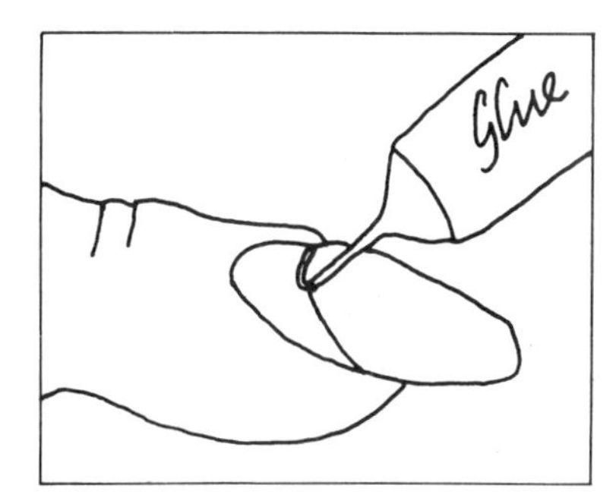

Re-apply glue to ridge

7 Hold firmly in position until both are securely bonded. (Orange sticks or plastic film can prevent client and operator being bonded together.) Ensure that the extension is parallel to the natural nail and does not bend over at the fingertip.
8 Apply glue to the top surface of the join.
9 Continue until all tips are in position.
10 Select an appropriate 'buffer'.

Note

Care and experience are needed for correct buffing of nail extensions. Although the term 'buffer' is used the tools have more in common with emery boards. They vary in both shape and texture. Square, round and spatula shapes are available. There are very smooth textures to shine and finish nails and very coarse which are needed for shaping plastic, nylon, or acrylic nails.

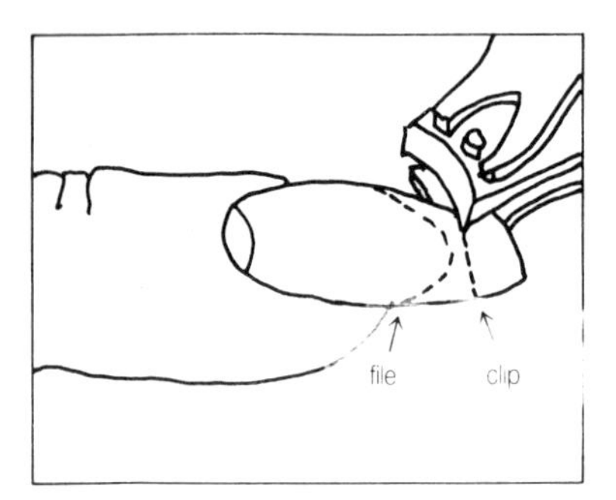

Use clippers to achieve realistic length

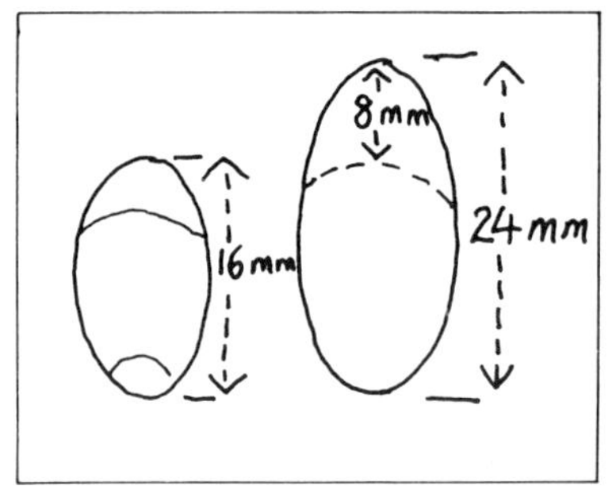

Maximum extention length—half of natural nail

11 Buff the join with a medium buffer. Make sure the natural nail does not become damaged.
12 Do not remove the dust as this helps fill the ridge.
13 Apply glue to the complete surface of the nail and allow to dry thoroughly – this is essential.
14 Run a finger along the nail surface. If the nail tip and natural nail join can still be felt repeat the buffing and gluing (steps 11, 12, 13), until a smooth join is achieved. As many as four applications of glue may be needed.
15 Clip the nail tip so it is slightly longer than required. Use clippers or scissors taking care that clippings do not fly into the eyes, either the client's or the operator's. Advise the client on a realistic length. Overlong extensions often accidentally come off. The recommended maximum length for an extension is about half the length of the natural nail.
16 Use a coarser file or buffer to shape the free edge to the desired shape.
17 Turn the hand over and apply a drop of glue between the tip and the natural nail. This prevents water, debris, or bacteria lodging in the join.
18 Apply cuticle oil around cuticle to counteract any drying due to buffing.
19 Clean the nail surface with 70 per cent alcohol and proceed to polish application (see p. 31).

Safety Tip

Take care when cutting – nail chips can fly into the eyes.

Note

Extension plastic tips must be removed, or repaired at the cuticle end after about three weeks as the natural nail grows.

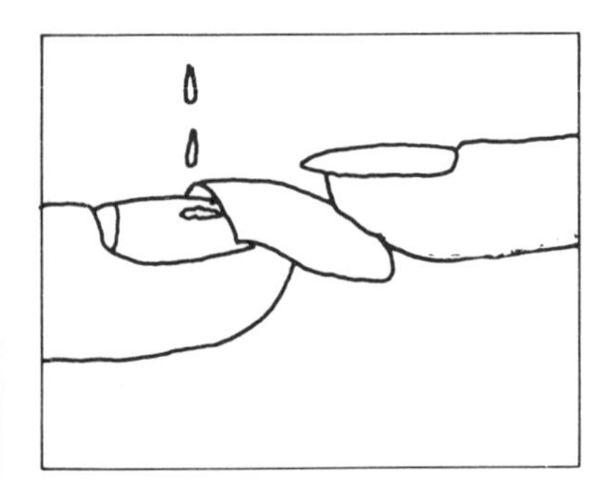

Removal. Apply solvent—gently rock

Nail Tip Application with Gel and Light Systems

There is a distinct advantage in using the gel and light system when applying nail tips. The gel is self-levelling and needs less buffing and gluing to achieve a smooth join. A fuller consideration of the uses of gel and light systems is given in Unit 10.

Procedure

1 Select and apply the tip with instant nail glue. (See p. 62, steps 2–9.)
2 Clip and file the nail tip to shape. (See p. 64, steps 15–16.) Take care that the sides of the tip, where it meets the natural nail, are given particular attention.
3 Remove any filing dust/debris from under the tip, nail, and free edge, with a brush.
4 Clean the nails with an antifungal cleanser and gauze. Do not use cottonwool which could leave fibres on the nail.
5 Apply natural nail primer – sparingly – to the natural nail only. Allow to dry completely. This is an important step. If too much primer is used or if it is not completely dry, the gel will not adhere to the nail properly.
6 Apply nail tip primer, again sparingly, to the top and underside of the nail tip.
7 Apply gel over the complete nail as described in Unit 10 (p. 68).
8 Cure the nails with ultraviolet light according to the manufacturer's instructions.
9 Apply a more generous second coat of gel. Make sure the brush is brought over the free edge of the nail to the underside of the nail tip.
10 Cure the gel. Do not forget to turn the hand over to cure the gel under the free edge.
11 Cleanse and check the nails as described on p. 69.

UNIT 10

Gel and Light Systems

Gel and light systems have developed from technology used in dentistry. They consist basically of self-levelling gels that are 'cured' (hardened) in minutes by exposure to ultraviolet light. A wide range of ultraviolet machines are available from different manufacturers. They all operate on similar principles.

A recent appearance on the market are gels that need an accelerator spray instead of ultraviolet. The danger is the considerable heat that is generated by the hardening spray. This can be extremely uncomfortable for the client.

Gel and light systems can be used in a variety of ways to meet individuals needs.

COVERING THE NATURAL NAILS

Gel can be used to strengthen and protect the natural nails, or to disguise layering.

PROCEDURE

1 Complete the manicure up to the polish application. Make sure all the dead cuticle is removed from the nail plate.
2 Lightly roughen the surface of the natural nail with a medium buffer/file. This allows the gel to form a perfect adhesion with the nail plate.

Care Note

The temperature of the salon, and the client's hands, is critical to the success of bonding: 21–27°C (70–80°F) is considered ideal.

3 Thoroughly clean away filings with a brush or gauze. Do not use cottonwool which may leave traces of fluff on the nail.
4 Clean the nails thoroughly with a non-oily polish remover, using gauze, not cottonwool. Most manufacturers market specific cleansers for this but either acetone or alcohol (70 per cent) will do equally well.
5 Apply a disinfectant to the natural nail plate. Use the product recommended by the manufacturer.

Care Note

This step is very important. It stops bacteria and fungi multiplying between the gel layer and the nail.

6 Use a natural nail primer. This is a critical step. If too much is used it will not dry properly, causing poor adhesion between the gel and the nail. The primer should cover the whole nail plate evenly, including the sides and bottom, for effective adhesion.

Safety Tip

Most primers contain methyl methacrylic acid which etches (slightly roughens) the nail plate. This must be used sparingly. If any accidently touches the skin or clothing, immerse in cold water immediately, then wash with a mild soap.

7 Allow the primer to dry completely, as shown by a white, chalky colour. If the primer is not totally dry it will not bond with the gel.
8 The order in which gel is applied is determined by the ultraviolet light system used. In some systems, all the nails of both hands have to be cured together. Other systems require each nail to be cured separately, and others cure two or three nails at one time. The manicurist must follow the manufacturer's recommendations.
9 Apply gel to the nail plate. Pay particular attention to the

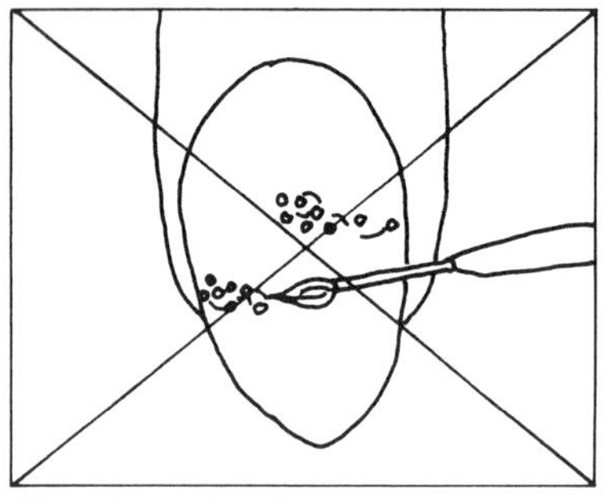
Do not overbrush

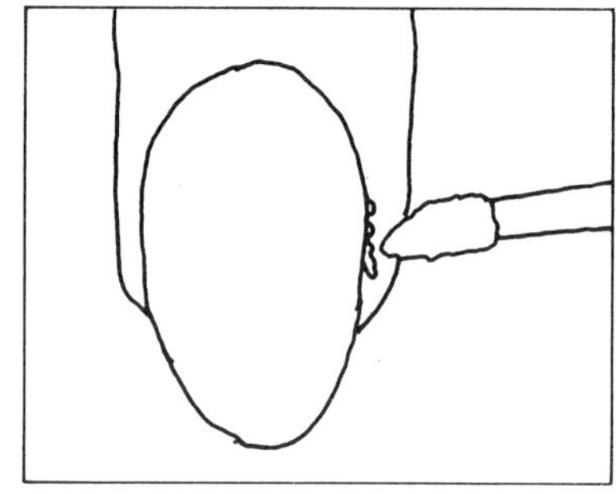
Remove any excess from skin

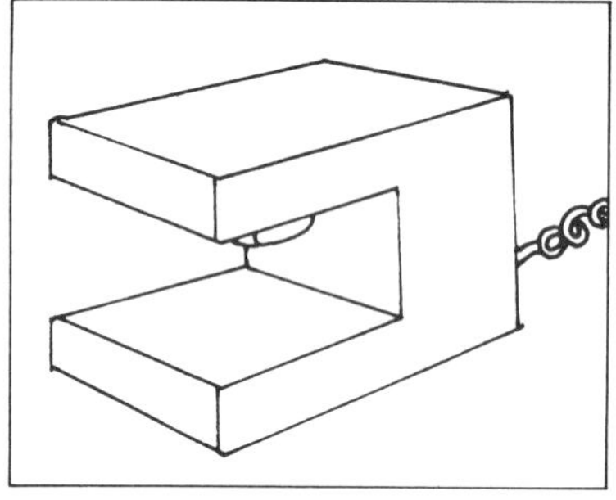
A gel and light machine

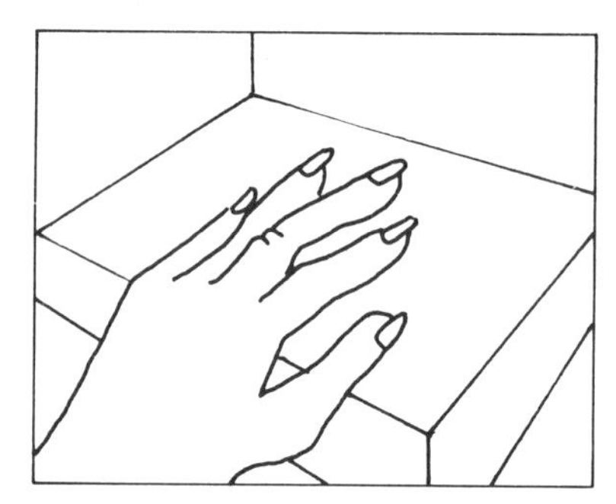
Hand position

sides and cuticle area. It will tend to shrink slightly when cured.

10 Use the type of brush recommended by the manufacturer. Some insist on natural sable, others prefer synthetic bristles. It is important not to overbrush the gel, as this may cause uneveness and bubbling. The gel is self-levelling.

11 Remove any gel accidentally overlapped on to the skin or cuticles with an orange stick before curing.

12 Cure the nails under the ultraviolet light according to the manufacturer's instructions. This usually takes about one minute.

Safety Tip

Don't panic if gel is accidently cured on the skin. It will lift easily once the skin's natural secretions have produced an oily layer between the skin and the gel. Removing it by force may cause damage and pain. It is difficult to limit special removers to small areas.

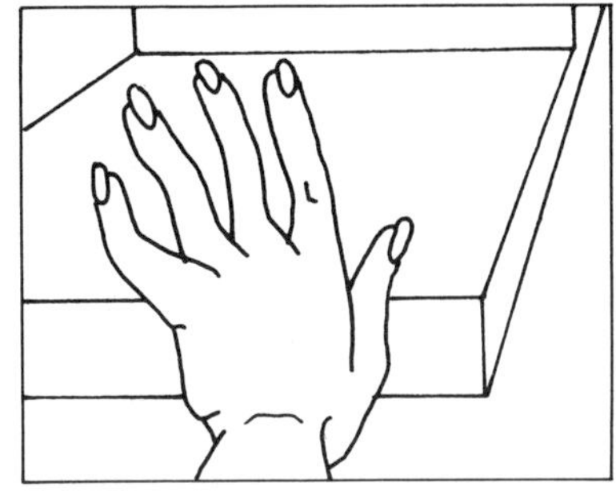

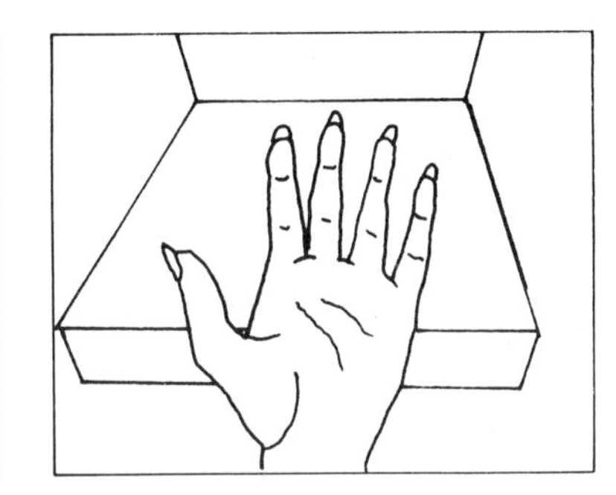

Turn hand over

Care Note

If the cleansing movements are too rough the surface of the nail will not be smooth, producing a dull rather than a shiny finish.

13 Apply a second coat of gel over the complete nail. Particular attention should be paid to the sides and base. Allow the brush to cover the free edge to protect from chipping and damage.
14 Cure as before but with double the exposure time. Make sure the sides have been properly cured by rotating the hand position. Turn the hand over to cure gel on the underside of the free edge.
15 Cleanse the nails with specially prepared cleanser, using gauze, not cottonwool, to avoid fibre residue. The correct technique is learnt with practice and experience. Excess gel is removed by first patting and then gently wiping the nail plate.
16 Check the nails carefully. If there are any bare parts repeat all the relevant steps.

ALTERNATIVE GEL AND LIGHT USES

The gel system can be used to form marginal extensions to the natural nails. This is particularly useful when the nails have been so badly bitten that there is not enough nail for any other extension to stick to properly.

To form these gel extensions, follow instructions for the gel overlay on natural nails (see p. 66) along with nail sculpture instructions (see p. 77).

The client must return within one week to check that the extended nails have not been damaged by repeated attempts at biting, and to repair them if necessary (see fill-ins and repairs p. 71).

This procedure is as important psychologically as it is physically, because it means the client has taken a step towards stopping nail biting. The client will need encouragement from the manicurist to help break the habit by using this treatment.

Continue to see the client each week, repairing and filling as necessary, until the natural nail has a visible free edge. At this point all the options for nail extension are 'at the client's fingertips'. One likely choice is for removal of overlays and sculpting. The client should be advised that the natural nail may be softened by the gel overlay/sculpting and will need 48 hours to harden. This is a danger point. It is best to remove the gel gradually, finger by finger, over two weeks to make sure the nail-biting habit has finally been broken.

Regular manicures are strongly recommended at this stage. The client should be advised to keep the nails short until the nail and the cuticle have recovered from perhaps years of biting.

Gel, Light and Complete Nail Forms

Gel and light can be used to overlay complete nail forms. The only significant benefit is that the overlay ensures a complete join at the cuticle, the sides and beneath the free edge. This marginally extends the life of complete nail forms, though they will of course need 'filling in' (see p. 71) in about three weeks due to natural nail growth.

Unit 11

Maintenance and Repair of Extensions (Fill-ins)

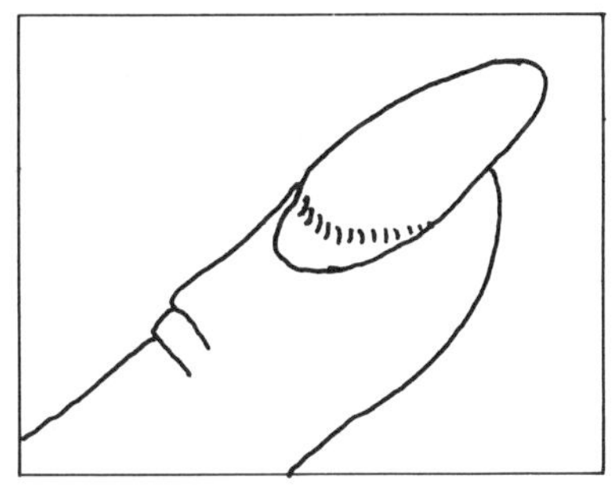

Growth after 2–3 weeks

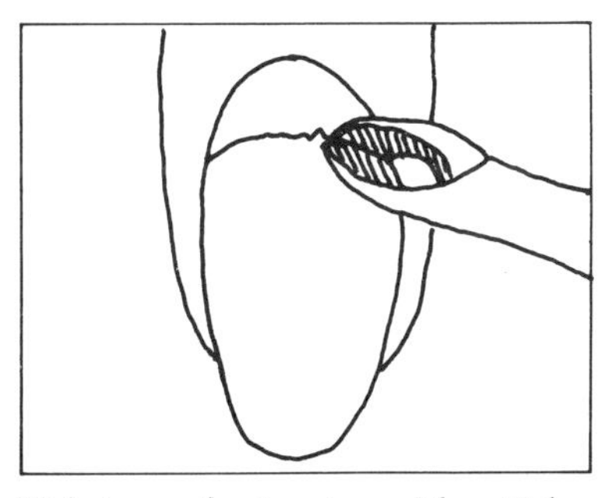

Tidy base of extension with cuticle nippers

> **Safety Tip**
>
> Eye protection is needed when clipping. Chips can fly and damage eyes.

Extensions need to be checked every 2–3 weeks for splits, chips and cracks which will need repairing. If the extension is to be maintained the client will need a fill-in between the cuticle and extension every 2–3 weeks – depending on the growth rate.

It is important that this is explained to every client **before** fitting extensions for the first time. If a sizeable gap is allowed to develop between the cuticle and extension, this will allow water, bacteria, fungal spores and debris to penetrate between the natural nail plate and the extension. This could cause fungus growth, for instance, which could permanently damage the nails. If the client does not return after three weeks, it is good business practice to send a reminder card. Offer to remove the extensions rather than chance any damage (see removal of nail extensions p. 59). This will also protect your reputation – clients are likely to blame the manicurist, rather than their own neglect, if things should go wrong.

The method of repair and maintenance is similar with all types of extensions. It is better to use the original products – gel, acrylic, glass glaze – being careful to follow the manufacturer's instructions for each particular system.

Procedure for Maintenance and Repair

1 Remove any polish with a non-acetone polish remover.
2 Clip any extension material that is loose or lifting from the base and around the cuticles. Use very sharp cuticle nippers but do not dig or pull as this could be uncomfortable for the client.

3 Complete cuticle treatment as for a normal manicure – cuticle massage, soak, cuticle removal techniques (see p. 27).
4 Remove the shine from the natural nail with a buffer or emery board taking great care not to damage or tear the cuticle.
5 The technique will vary at this point depending on the type of extension used. Refer to the original instructions and use the products in exactly the same way.

Note

Some manicurists prefer to re-glue around the edges before proceeding. Others find this makes for greater difficulty in producing a thin and flawless join at the fill-in line. Experiment with the products to be used. This is best done on models – not clients – until the technique is perfected.

The final result should be exactly like the original extension – thin, perfectly smooth and as close to the cuticle as possible.

Chipped or broken nails are easy to repair. Either remove and replace the complete extension, or use nail sculpture techniques (see p. 77).

GLASS GLAZE TIPS

Glass glaze tips protect or extend nails without using ultraviolet light hardening. The process can be used in conjunction with a glass fibre mesh and a hardening spray, which produces a very natural-looking finish without the need for a coloured varnish protection.

When used with added nail tips, the technique produces extremely natural-looking nail extensions.

The **advantages** of glass glaze tips are:

- Application takes less time than with other types of extension.
- The extensions can be extremely light and thin.
- Because the extensions are not normally applied closely to the cuticle 'filling-in' is not usually required.

The **disadvantage** is that an impatient manicurist may apply the hardening and developing spray more frequently or generously than is recommended. This causes pain or discomfort for the client, arising from mild to serious heating of the nail plate. In severe cases the nail bed can be permanently damaged by the chemical reaction.

Procedure

Follow the procedure for attaching tips with nail glue – cutting, clipping and buffing (see p. 64).

1. Shape the nails.
2. Take the glass fibre mesh (silk or linen can be used instead) and cut a piece that fits from the cuticle to the free edge, and side to side across the nail.
3. Gather the material slightly at the cuticle end. Place a drop of instant nail glue on the nail, near the cuticle, but taking care to avoid the cuticle.

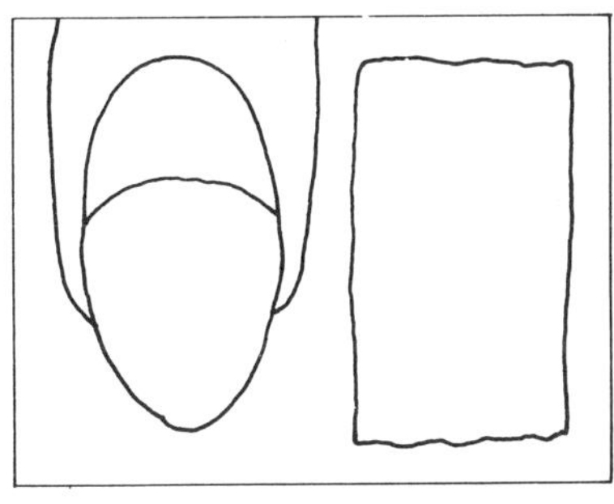

Cut fabric to length and width of nail

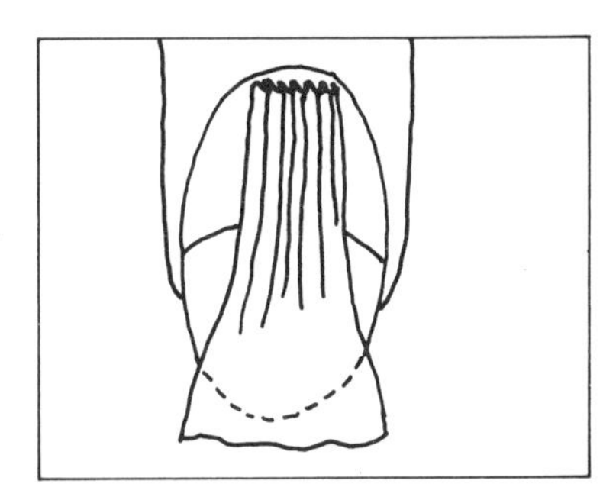

Gather fabric at cuticle end

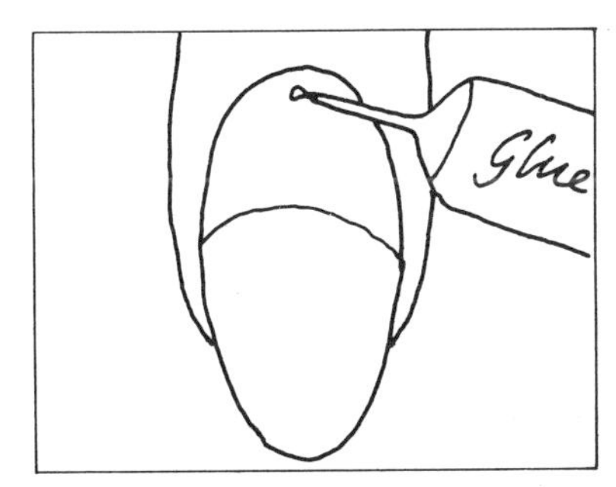

Glue close to cuticle

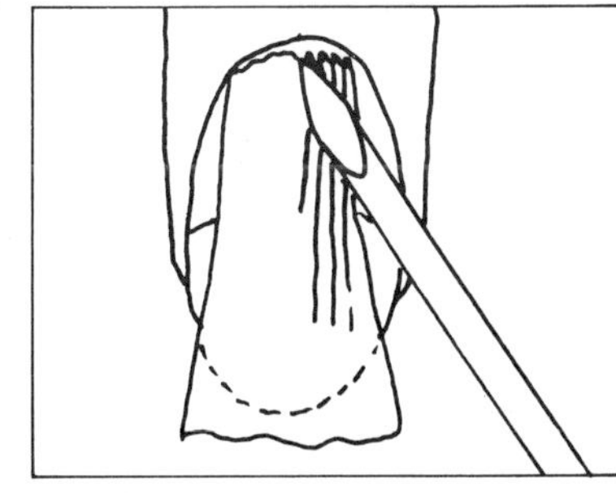

Spread out with orange stick

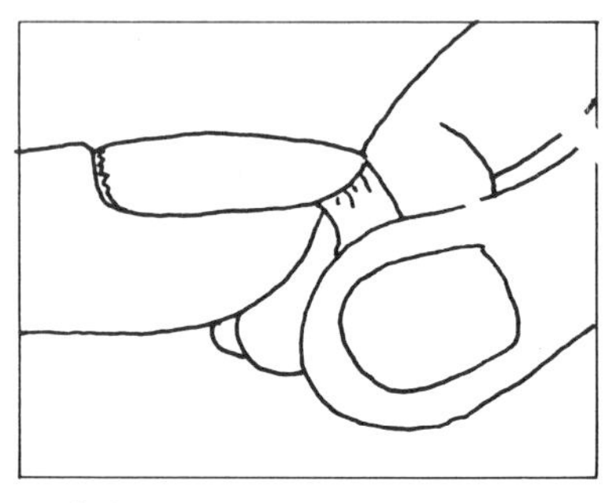

Lightly stretch towards free edge

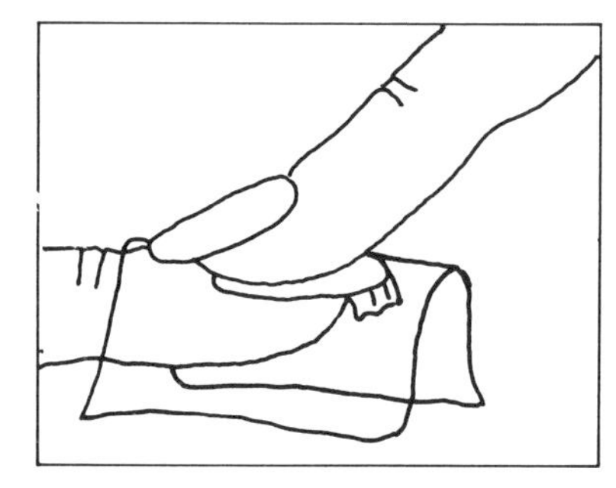

Press down using plastic film

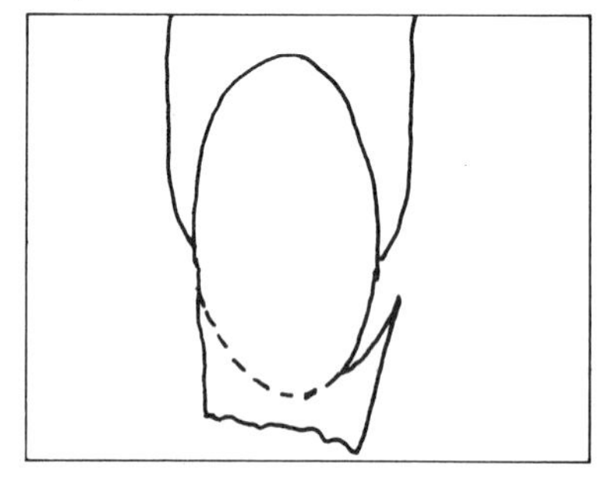

Snip away excess fabric

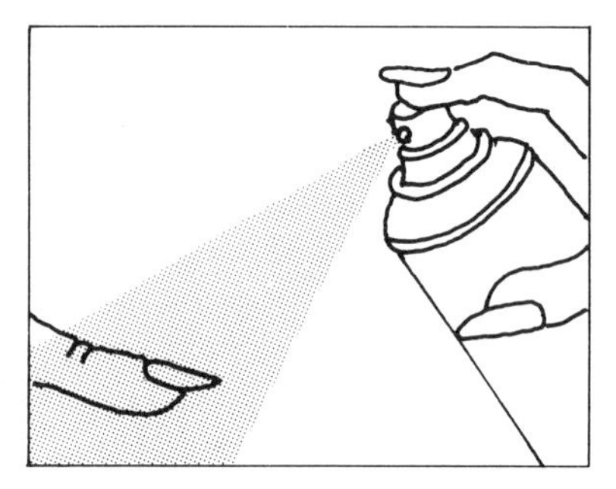

Apply hardening spray

4. Slide the fabric into place. Spread it out on the nail plate using an orangewood stick dipped in solvent.
5. Lightly stretch the fabric towards the free edge. Cover the entire nail plate and the join of the extension. Re-apply glue over the surface.
6. Press the fabric down – without becoming stuck to the client. Use cling film to avoid contact with the glue. The pressing action removes any air bubbles and creates a smooth bond.
7. When the glue is dry, snip away the excess fabric and file the free edge with an emery board.
8. Re-apply glue if necessary.
9. With an emery disc, buff the surface of the nail until it feels silky smooth.
10. Apply glass glaze evenly over the complete surface. Use the drying accelerator spray at the recommended distance (see dangers p. 38). Repeat up to four times allowing the glaze to dry between applications.

Repairs, maintenance and removal

Follow the procedures described on p. 71.

Unit 12

Nail Sculpture

Nail sculpture is considered the ultimate test in the art of nail extensions. If a manicurist (nail technician) can produce a natural-looking nail using this method then other techniques become extremely simple.

The learner should expect to take three hours to complete the first full set and the finish is likely to be absolutely dreadful. It will probably take at least 30 full applications before a full set can be completed in under $1\frac{1}{2}$ hours – with an acceptable result – unless the manicurist has a lot of natural talent.

This should not discourage anyone from trying. It really does take a great deal of experience and practice. This skill must be mastered if a manicurist is to compete in today's profession.

The technique uses an acrlyic resin which is mixed on the brush immediately before application. A sable brush is dipped first into a liquid (the monomer) and then into a powder (the polymer). This is then applied to a nail form to extend the nail. It sounds easier than it really is.

Before starting there are a number of points to consider:

- See demonstrations of a variety of products before starting training or operating.
- Consider the working environment. Is there air conditioning or excellent ventilation to cope with a high level of chemical odour?
- Is there a clientele that will want this service? Is the capital outlay on training and stock worthwhile? Even if the do-it-yourself, trial-and-error method of learning is adopted there will still be the investment in materials and

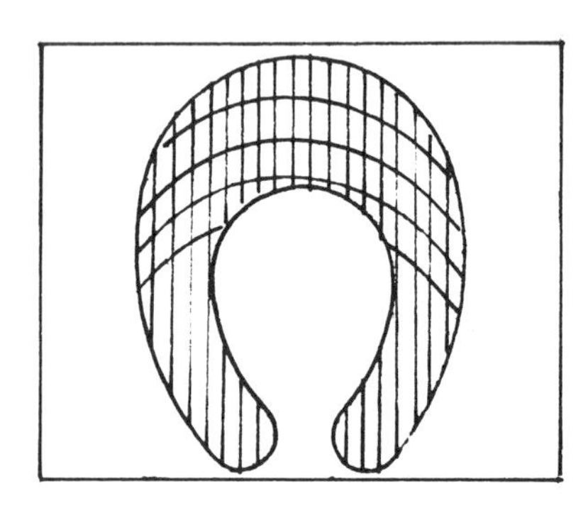

Disposable nail form, actual size — may be paper, foil or plastic

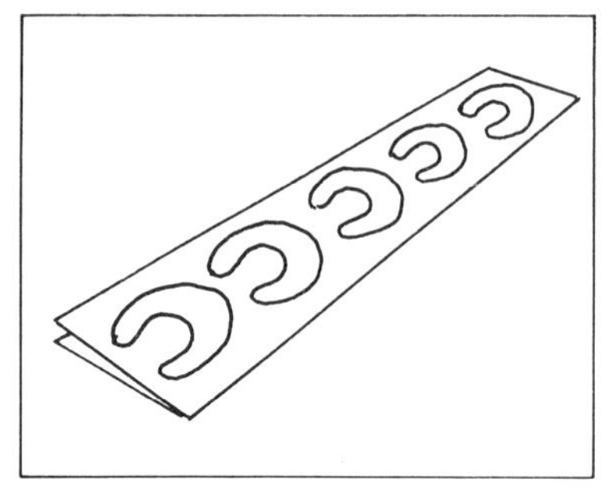

Folded strip of adhesive-backed disposable nail forms

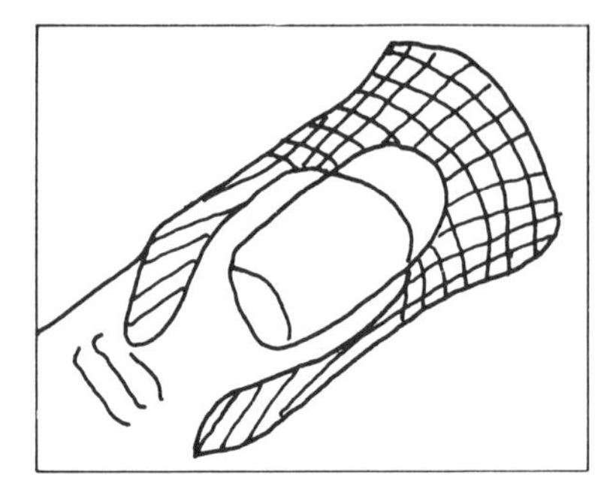

Disposable nail form in place

at least 100 hours of practice before attempting to start on clients.

- Look at hard nail forms – usually aluminium – that can be reused, and at the disposable ones using foil. Which will be easier to use? Perhaps you will need both for different purposes.
- You will need to invest in a wide variety of shape and size of sable brushes. Experiment will determine which kind will be the easier to use for different purposes.
- Is there a need for a variety of coloured powders – pink, natural, white or clear?
- Does one liquid evaporate at a faster or slower rate? This can be either an advantage or a disadvantage depending at the rate of working. Faster working means more appointments and hopefully more pleased clients. Slower working is easier and may produce better results.
- Are the liquids in bottle sizes appropriate for the number of clients to be attended? Too large bottles and too few clients creates an evaporation problem that will make application difficult.

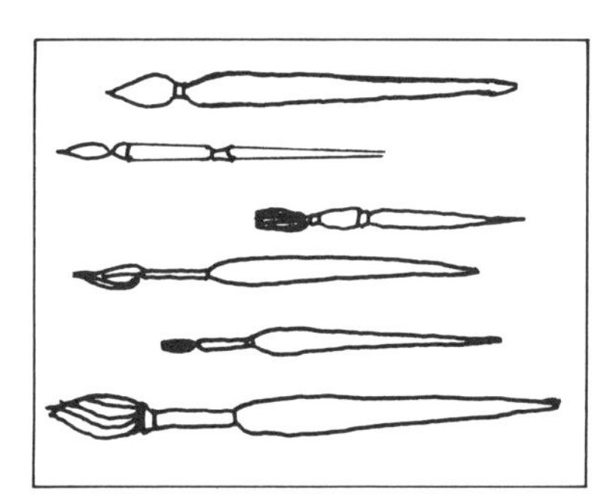

Range of sable brushes

These questions should be considered before entering the competitive world of nail sculpture. The additional skills can be very exciting and profitable.

This step in your training needs serious consideration. A lot of money is being made by training manicurists who in the end are unable to offer a competent service. There are franchises that promise a great deal but do not deliver the necessary level of training or support.

Hopefully, consideration of these points will not be a

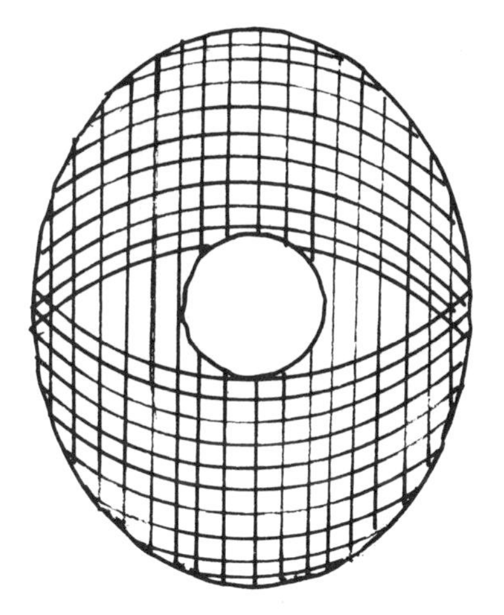

Aluminium templates—actual size

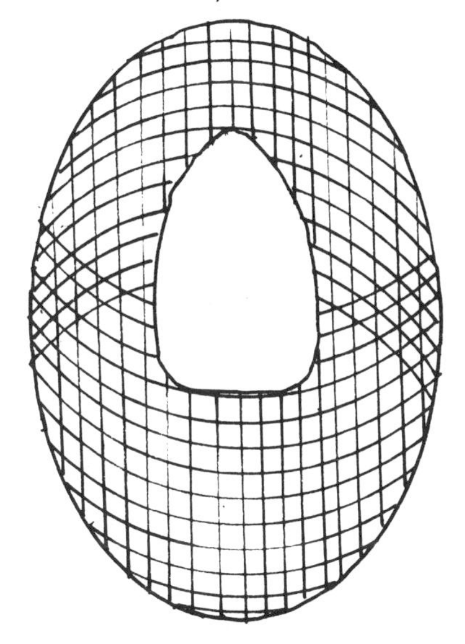

deterrent. There has to be the will to succeed and the best route is through constant practice.

Poor nail extension receives a lot of publicity through dissatisfied clients. Anything less than a competent and professional service damages the technique's image. If manicurists look carefully at all the facts before training then it can be hoped that only the highest standards will be achieved.

Procedure for Sculptured Nails

Preparation

1 Buff the natural nail surface with a fine emery buffer. Dust the nails clean with an antifungal treatment.

2 Choose an appropriate nail form. These may consist of the following:
   (a) Pre-formed aluminium, which can be fitted, formed and reused a number of times
   (b) Disposable horseshoe-shaped self-adhesive foil
   (c) Plastic horseshoe-shape

   The important factor is that the chosen template/mould should be of similar size and bevel to the natural nail, or adjusted to fit. If the template is not carefully chosen or formed, the resulting nail shape will not look natural. A pre-formed aluminium template can be bent and shaped to match the natural nail curvature exactly. If necessary a brush handle can be used to attain the required shape. If disposable horseshoe-shaped self-adhesive foil is being used, remove it from its backing and place it carefully – adhesive side down – on the finger. The

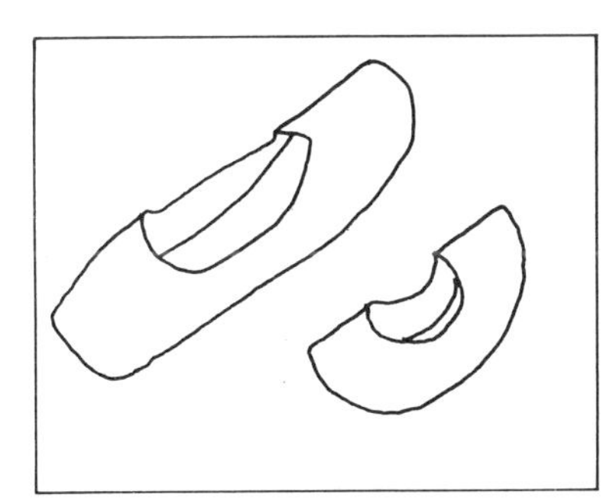

Bend aluminium templates to match curve of nail

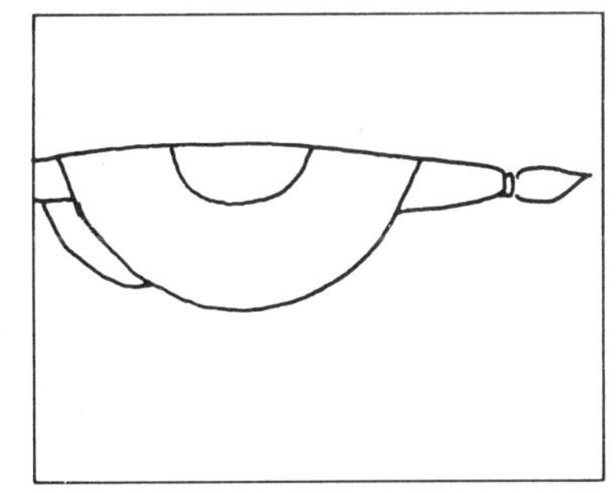

Roll over a brush handle or pencil

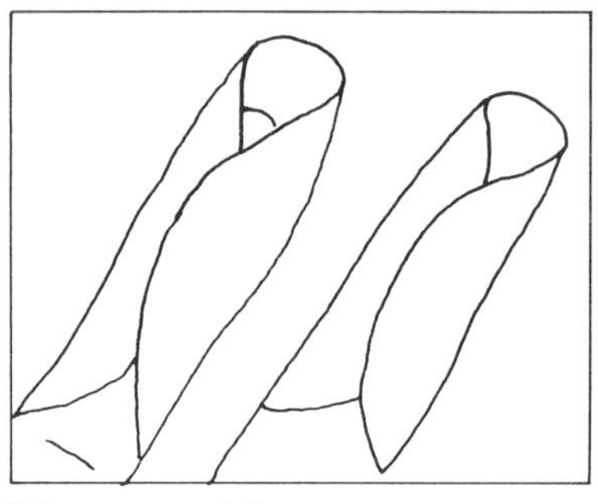

Wrap around finger

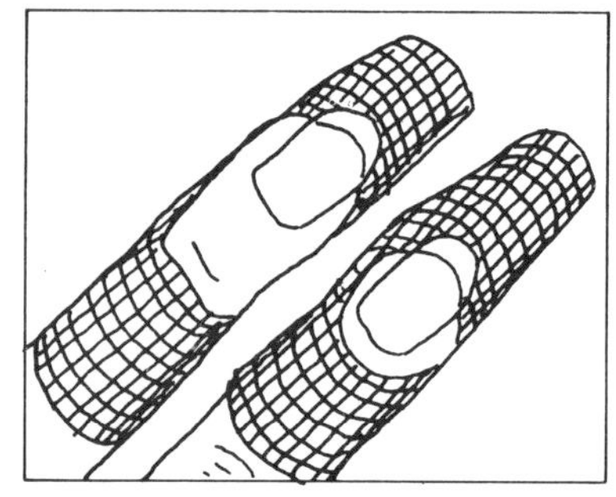

Templates in place—guide lines help shape new nail symmetrically

inside curve should be under the free edge and the outside edges, to maintain the curve under the free edge.

3 Carefully examine the line of the form. It should be in direct line with the natural nail. If it is not perfect try, gently, to alter its shape. An orange stick may be used to achieve this. If this is not satisfactory start over again. The nail form may have become loosened from under the free edge and will make the subsequent stages of procedure extremely difficult.

4 Place 14 g ($\frac{1}{2}$ oz) of polymer powder into a small glass dish and 14 ml ($\frac{1}{2}$ oz) monomer liquid (often methanacrylate) into another dish.
Different shades of powder are available. An experienced nail technician is able to produce an extremely realistic looking nail, with a pale-pink body and a white tip. This is achieved by a French manicure in

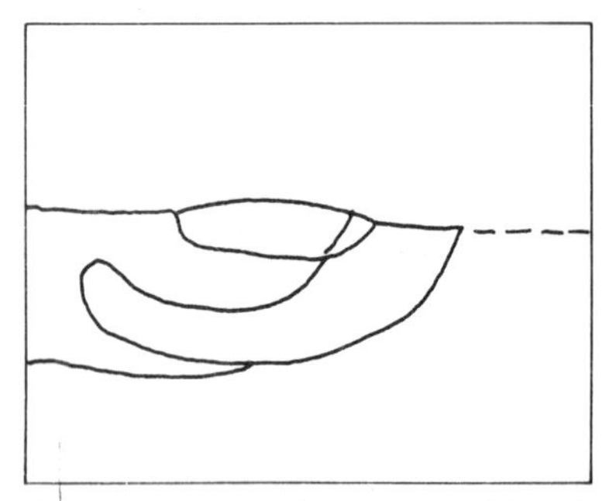

Template correctly positioned level with nail

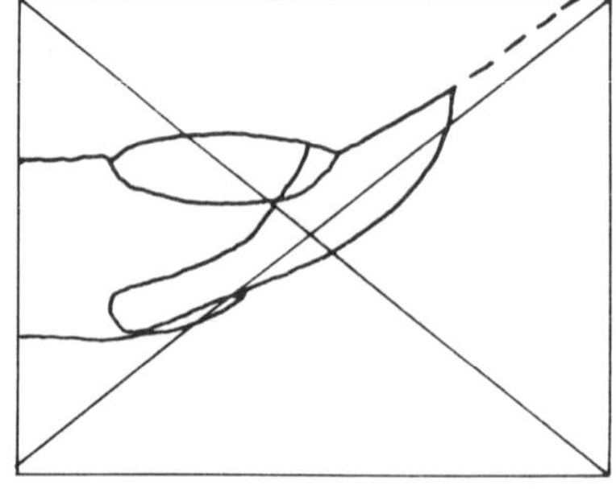

Sides positioned too far back make template curve forward

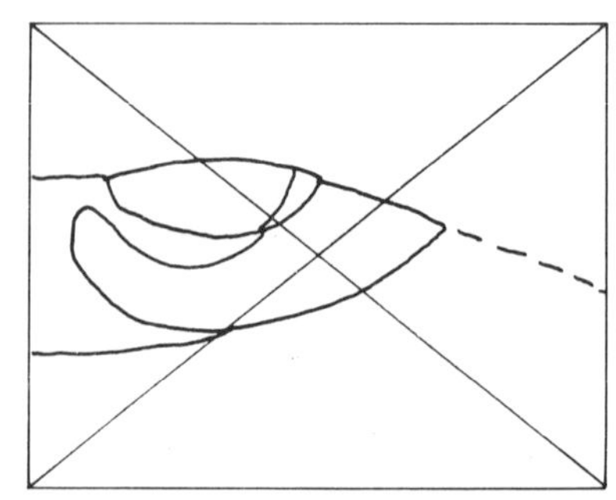

Sides positioned too far forward make template curve back

which different coloured powders are used to build different parts of the sculptured nail.

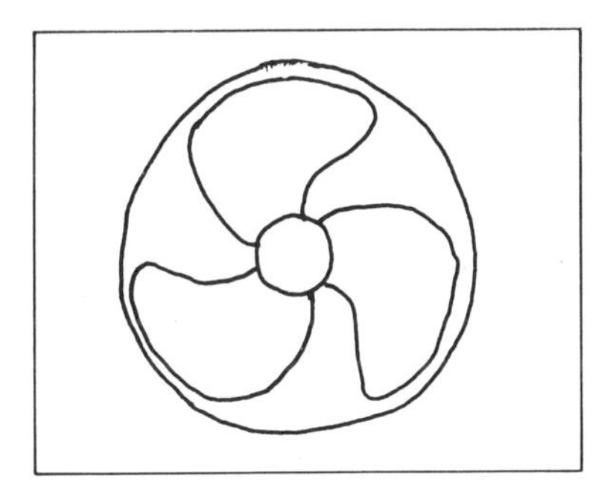

Toxic fumes—ventilation or extraction ESSENTIAL

Care Note

Take care – the fumes can be dangerous and unpleasant for both client and manicurist. **Use only in a well ventilated room**. An air-conditioned room with controlled temperature is ideal. If the room is too hot, nail sculpture becomes difficult due to rapid drying of the material. Higher temperatures also intensify the amount of fumes.
Odourless liquids and powders are becoming available but these are much more expensive, and adequate ventilation is still necessary to deal with the fumes produced.
Expect to use the liquid and powder in the ratio of 2:1. If the liquid has a high fume level, use a pot with a small hole for the brush. This helps to reduce evaporation. Always keep the liquid well covered when not in use.

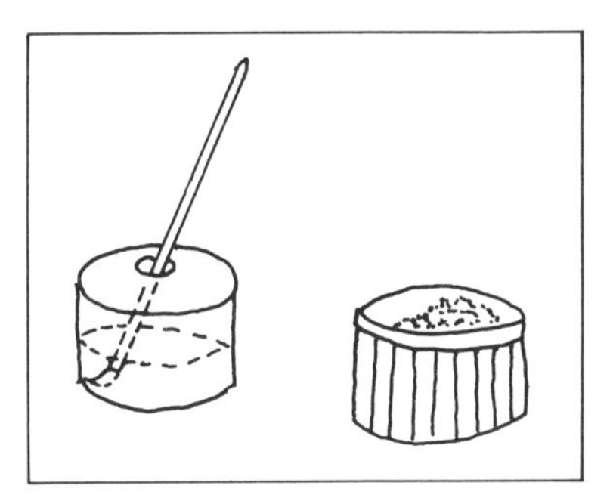

Dip brush in monomer liquid, removing excess, then in polymer powder—liquid: powder = 2:1

5 Dip the brush (see brushes p. 117), into the liquid and remove excess against the side of the dish. This is important for success. The correct amount of liquid to be used on the brush can be learned only through experience. Practise at first with the minimum amount possible. Further excess may be removed by touching the brush with gauze. If too much liquid is used, the paste will be runny and difficult to control.
6 Gently touch the brush on to the powder. Pick up as much as the brush will hold. Make this into a ball of paste on the tip of the brush.
7 Hold the template firmly in place with the fingers. Place the paste on the tip of the nail, where the nail plate and template join. Roll the brush until the ball of paste is in place. (Some manicurists prefer to begin the application at the base of the nail but it is easier to build the tip first.)
8 Use the brush to flatten the paste. Use small, gentle movements to create the artificial free edge. Aim for a thickness that is only a little more than a thick natural nail.
9 More liquid and powder will probably be needed at this point. Repeat steps 5 and 6 but make sure the brush is clean before dipping it into the liquid. Decide whether a white tip and a pink nail body are to be created. Mix the

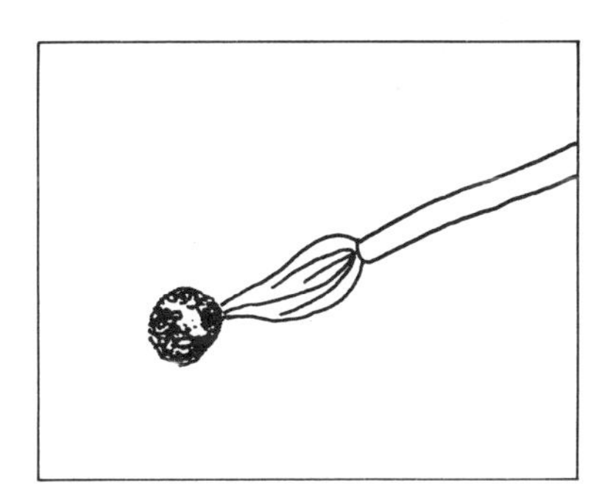

Ball of paste formed

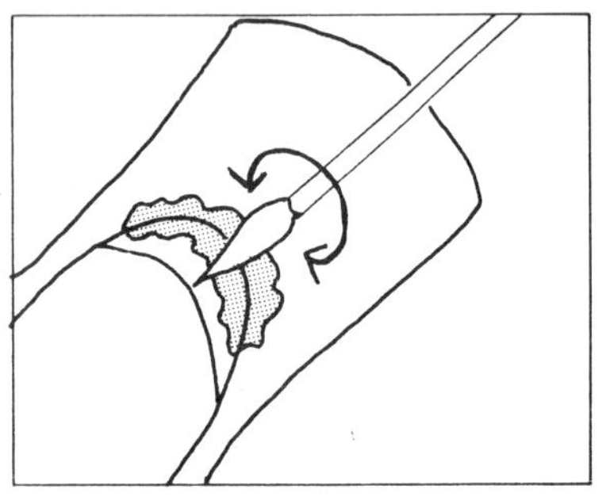

Roll brush to spread paste across free edge

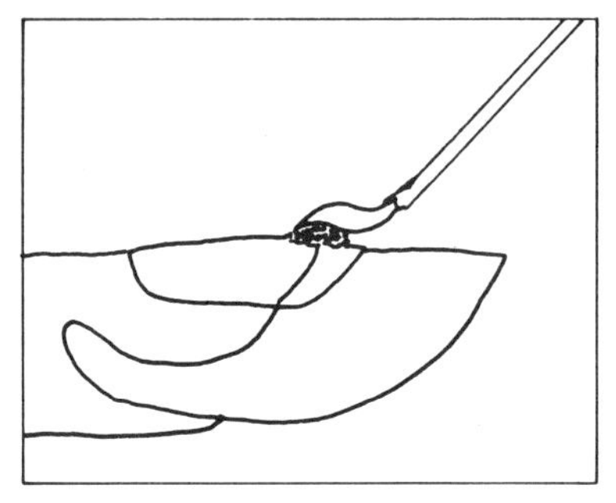

Flatten paste with brush

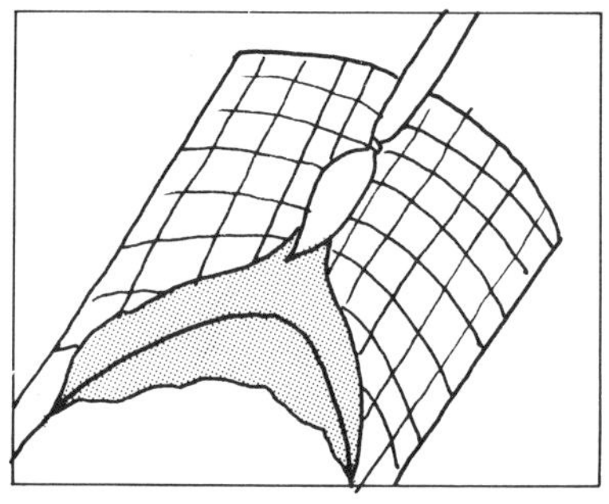

Small gentle movements create free edge

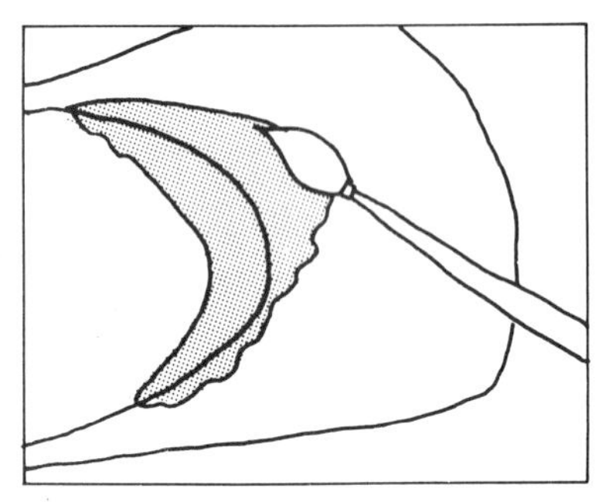

Build clean edge—use side of brush to pull paste towards tip

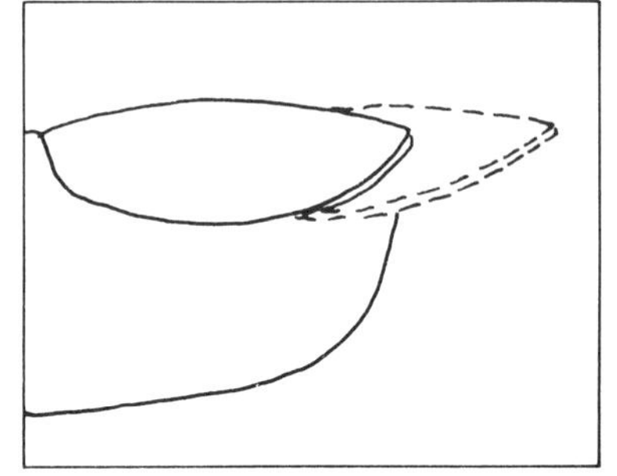

Thickness to aim for

liquid and powder as required. Take care not to extend the white mix below the flesh line of the natural nail.

10 Build a clean side of the nail tip by using the side of the brush to pull the paste towards the tip.

11 Continue to mix the liquid and powder of the chosen colour. Begin by building on to the nail plate to meet the free edge – as in step 8. Try to reduce the thickness of the paste application when working down towards the cuticle. This will help to create a natural-looking bevel.

12 When approaching the cuticle and side areas of the nail, reduce the amount of liquid and powder on the brush. If preferred a smaller brush may be used. Use the point to push the paste towards the walls and cuticle. Do not overlap.

13 Examine the shape and bevel of the nail from all angles. A greater bevel may be required. This may be achieved simply by placing more paste on to the previous application to create more of an arch. Check that the sides and cuticle area are adequately covered.

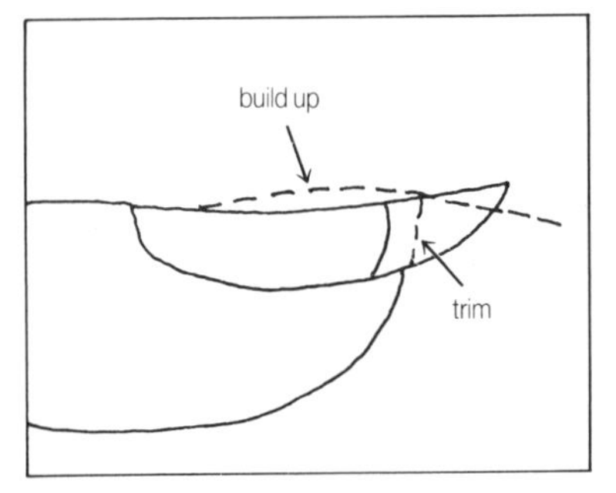

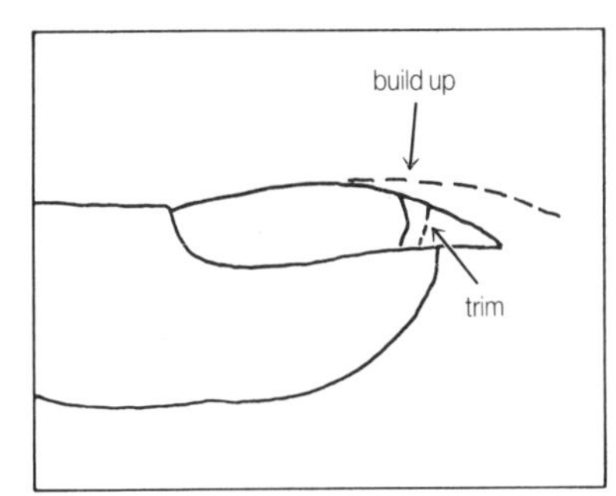

How to sculpt problem nail shapes

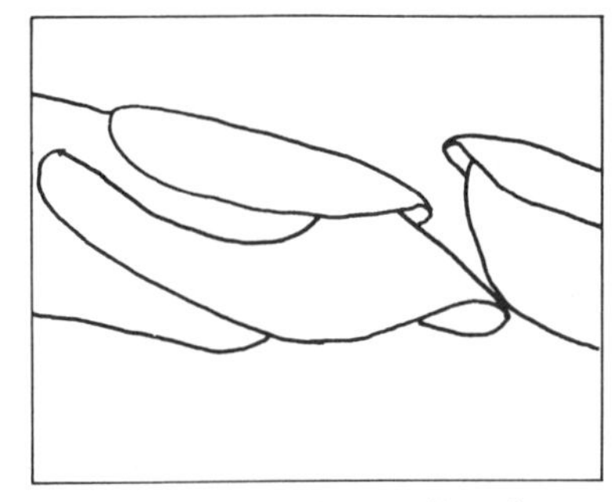

Remove template by bending down

14 Allow the nails to dry thoroughly. Carefully remove the former/template by bending it down away from the extension.

Filing

Filing is essential to produce natural-looking nails. This must be done carefully to avoid damaging the skin and cuticle.

The sculptured nail will be extremely hard to file. Specialist buffing/filing boards are available (see pp. 7 and 27). Some manicurists prefer to use electric filing machines but since the powders have become finer, and easier to file, machines are not really necessary.

When filing sculptured nails, all the manicure rules can be abandoned. The trainee may be shocked by the apparent roughness that the sculptured nail can be subjected to. The new nail is made from very hard material which would take too long to file using normal methods. A back and forth

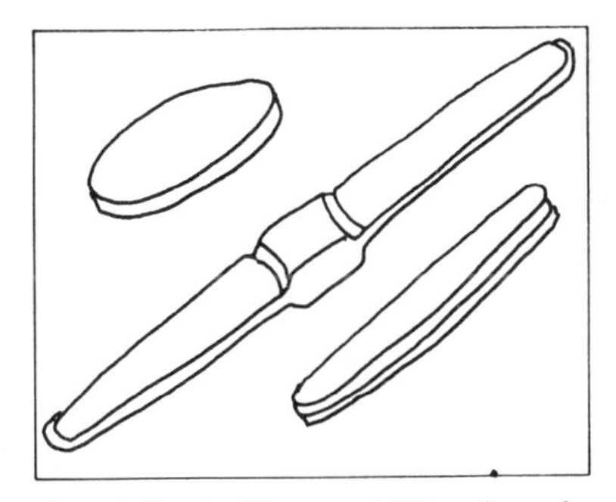

Specialist buffing and filing boards

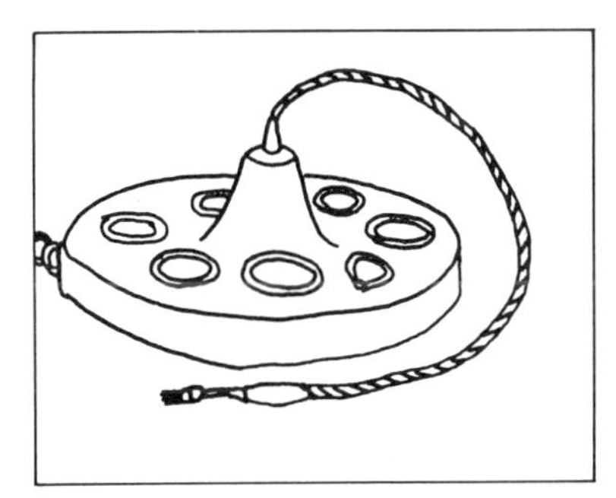

An electric filing machine

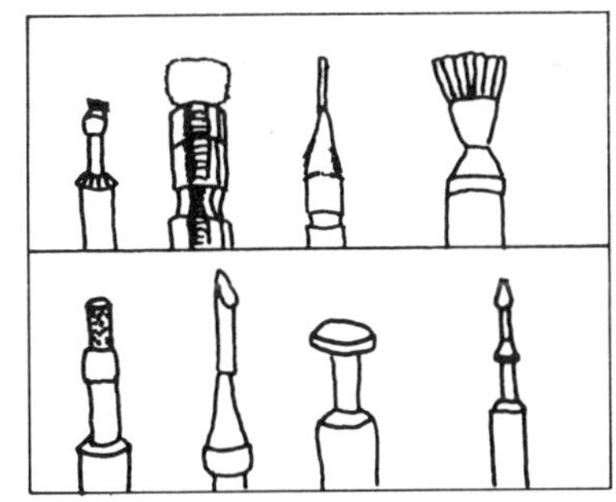

Attachments for filing machines—emery wheels, buffers, brushes, cuticle pushers etc.

sawing technique is needed. This of course must never be used when working on natural nails.

1 Check that the skin is clear of sculpture materials. If any remain, gently clip off as much as possible using nippers. Reassure clients that tiny amounts remaining will easily come off in a few hours as the skin produces moisture and oil.
2 Gently and carefully, file around the cuticle and sides of the nail to achieve a smooth effect.
3 File the centre of the nail to thin the nail plate until it looks natural. Work from the centre towards the free edge. Shape the sides from the centre out.
4 Shaping the sides and free edge takes a lot of time and care. Obvious excess length can be removed first with a rough file, taking care not to file too far. A normal emery board should be used to shape the sides. Pull the skin away from the sides to allow the thinner emery board to fit between the skin and nail. Use a normal manicure technique from side to centre of the nail to produce a natural-looking shape with no apparent join. Complete the free edge to the shape the client requires. Repeat this on all nails.
5 Check the underside of the nail to make sure it is neat and flat, with no rough edges. Remaining roughness should be filed as necessary.
6 Finish by polishing the nails to a smooth silky surface with a four-sided buffer (see pp. 23–4). Gently scrub the hands and nails in warm soapy water. Apply cuticle oil around the cuticles to improve the appearance.

The client should be advised on after care. Make sure the client understands that sculptured nails must be removed or 'filled in' professionally after three weeks (see maintenance and repair p. 71, fill-ins p. 71).

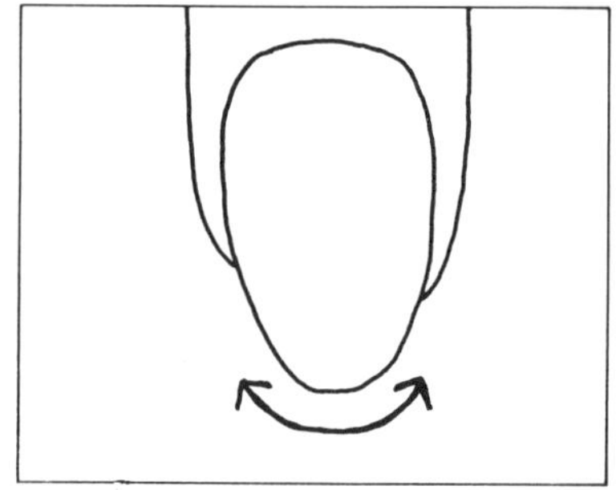

You can 'saw'—it's hard as nails

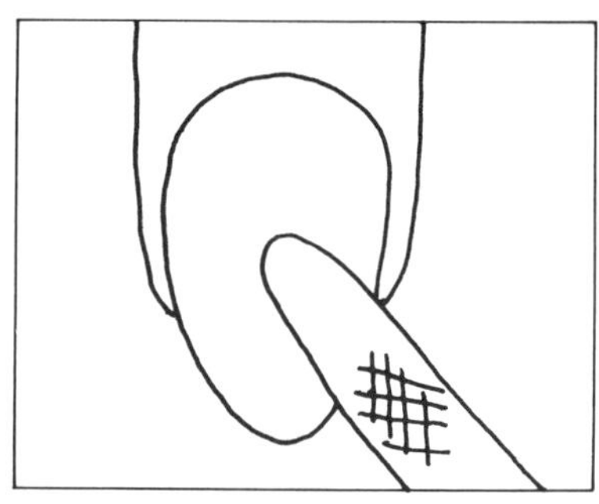

File to natural thickness from centre to free edge

UNIT 13

Nail Art

Some nail art designs

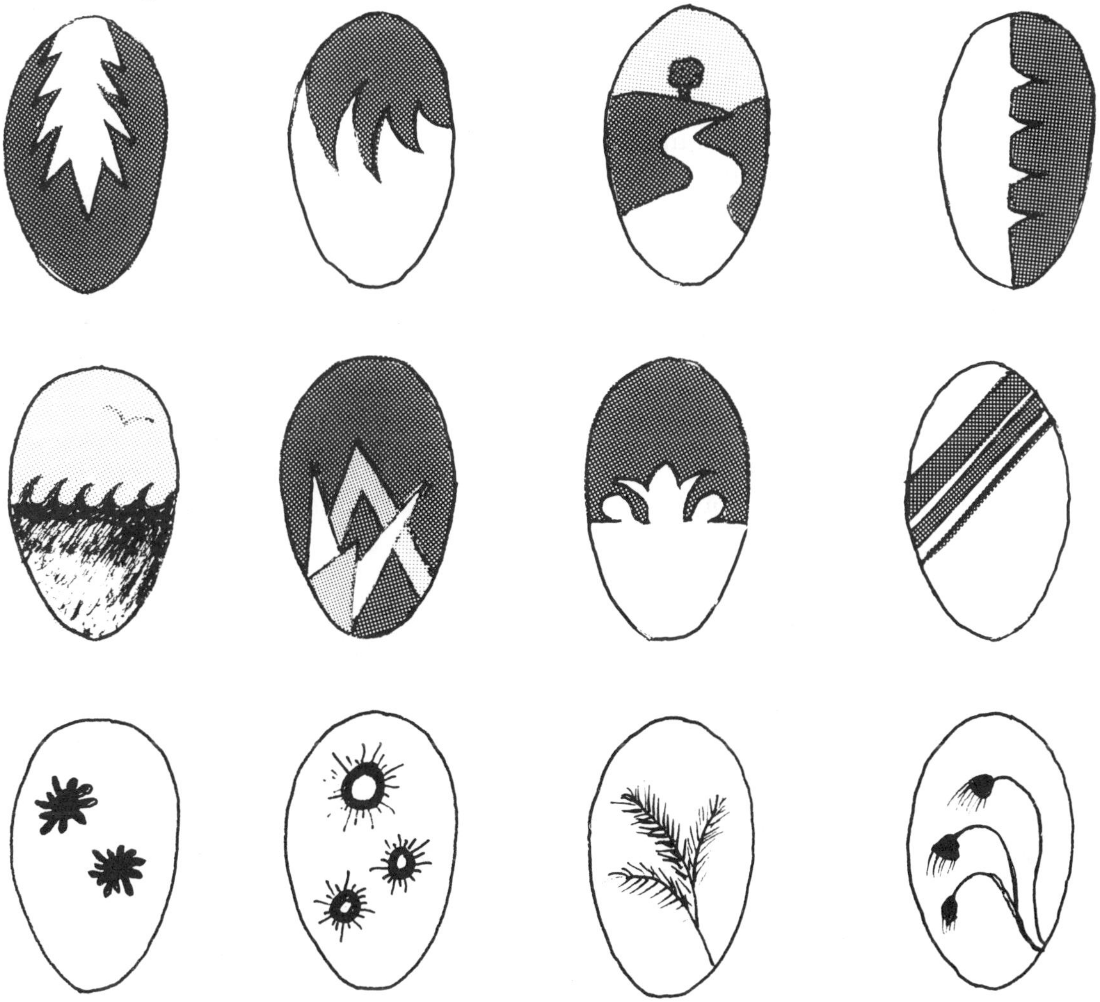

Nail art emerged in the USA in the 1980s and has become a developing trend. It started to gain popular appeal when a black, female, American runner – Florence Griffith Joiner, or 'Flo Jo' as the media dubbed her – won three gold medals in the 1988 Seoul Olympics. As Flo Jo came to the forefront of sports news, curiosity centred on her nails, which she used as exhibits of nail art. For the first time media attention worldwide was focused on nail art.

The trend has been adopted by manufacturers and 'instant nail art' is now available in a variety of forms. The following are two popular methods:

- Pre-prepared stick-on nails in gold or silver, or patterned with inset diamonds, chains, etc.
- Self-adhesive striping tape used to stick on top of coloured varnish in a wide range of patterns. It does not require a final top coat and can easily be removed in the normal way. It is very attractive for special occasions.

Adventurous manicurists may want to paint their own designs. The following may give some ideas – all that is needed is some masking tape and a few fine brushes:

- Experiment first on artificial nails. Choose a base colour and let the nail dry. Cut notches, curves, or zig-zags in the edge of a piece of masking tape and stick it over the part of the nail that is to remain in the base colour. Then paint a different colour over the tape, wait until dry, and peel the tape off.
- Different brush sizes make stripes of varying widths. Blobs, lines or circles can be 'feathered' by pulling the polish outward with a fine point, or a brush. Try a toothpick, a pin, a fine sable brush. Try the same techniques with a polish of a thicker, or thinner,

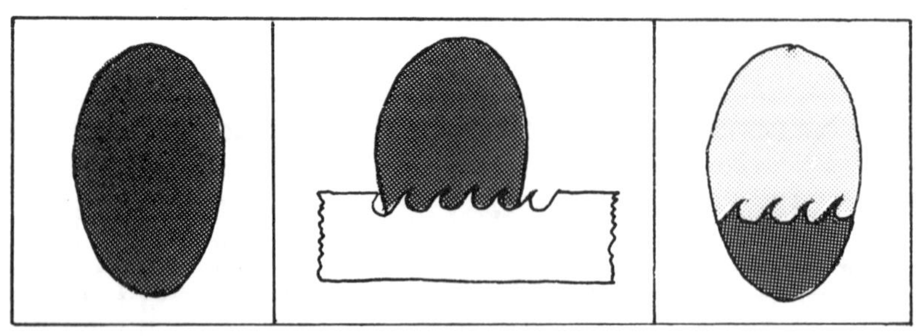

Paint base colour, leave to dry | *Stick on notched tape, paint second colour* | *Finished design*

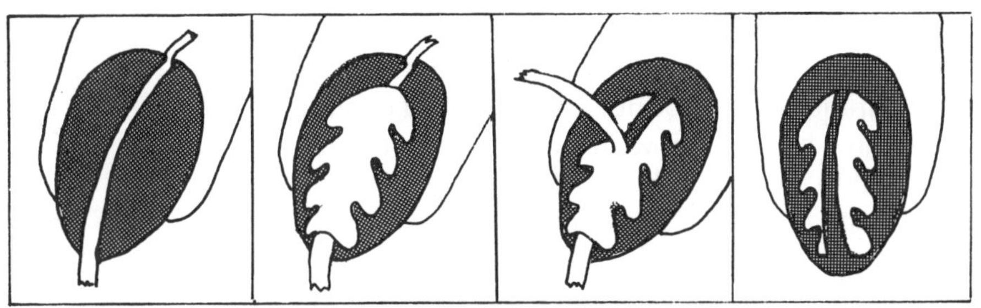

Stick tape over base coat | *Paint blobby shape over it* | *Peel off tape when dry* | *Finished design*

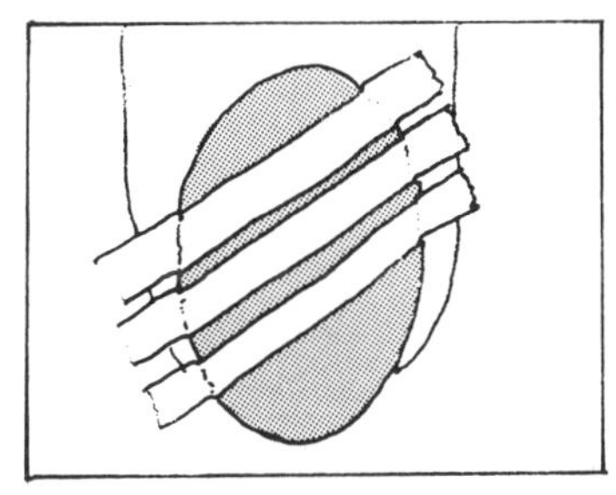

To make stripes

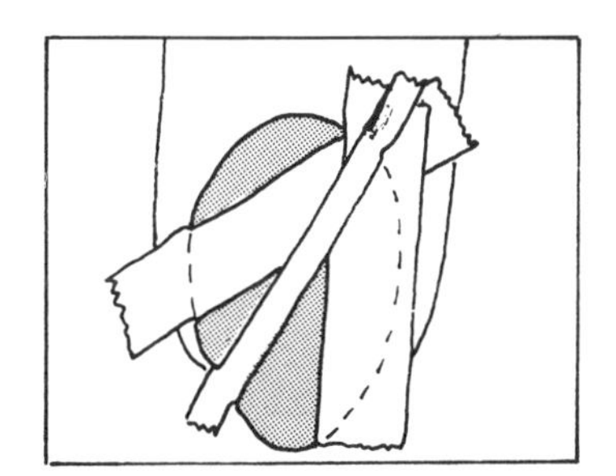

To make V-shapes

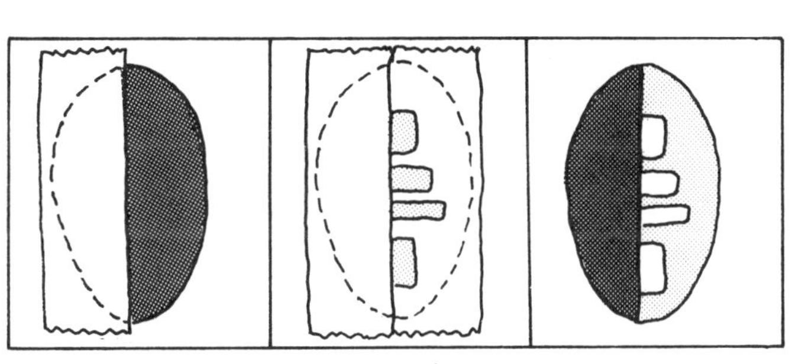

Paint second colour | *Add notched tape, paint third colour* | *Finished design*

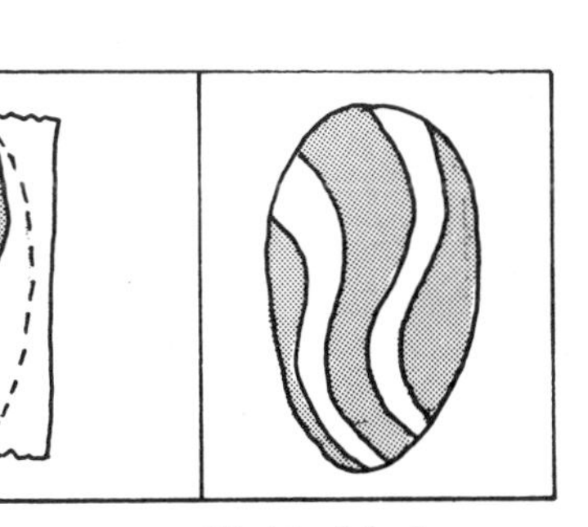

Cut curved lines | *Spread cuts apart* | *Finished design*

consistency. It will be found that flowers, leaves, stars and ferns can be made. A combination of these techniques can produce sophisticated designs. Protect the finished design with a coat of sealer, and advise the client to apply another coat after a few days.

- If the manicurist decides to specialise in nail art, an airbrush may be useful. This sprays acrylic paint, giving fine graduations of colour. Three-dimensional designs may be achieved with stuck-on beads, feathers, wire – there is no limit but the imagination.

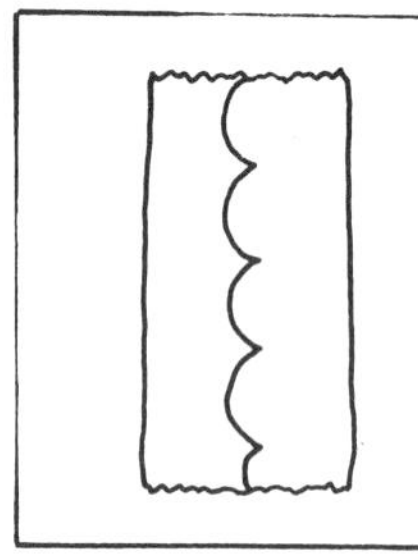

Cut curved line

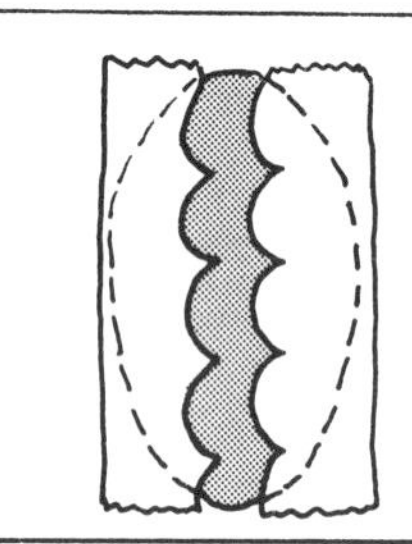

Spread tape sections apart

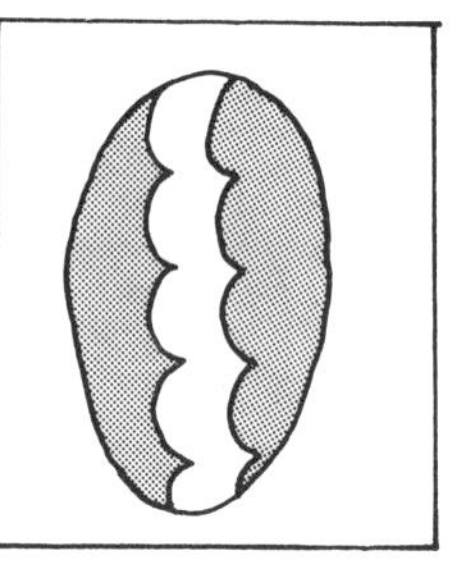

Finished design

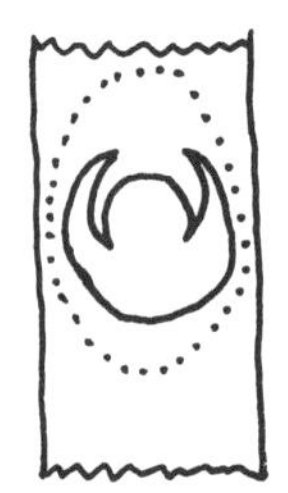

'Stencils' cut in tape—use a sharp blade

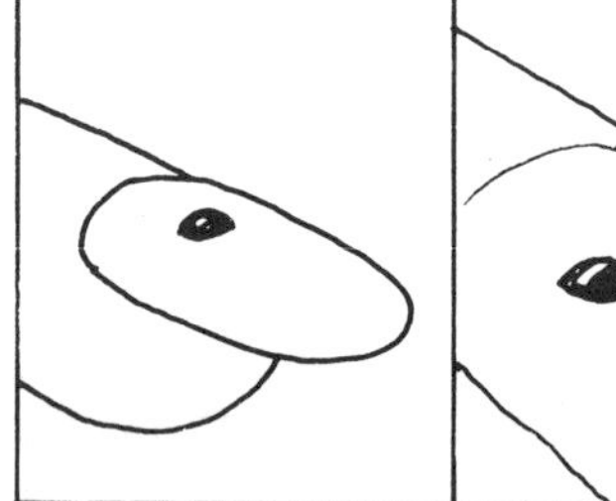

Paint blob

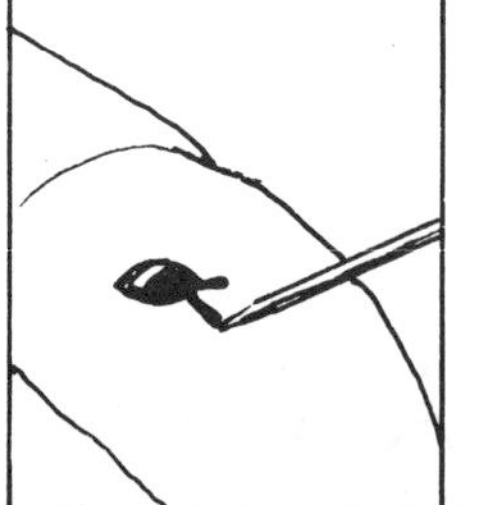

'Feather' by pulling outwards

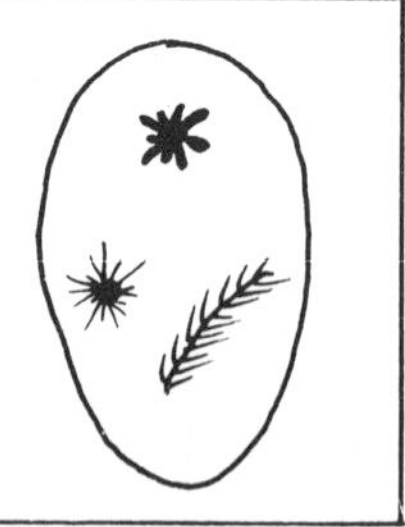

Some feathered designs

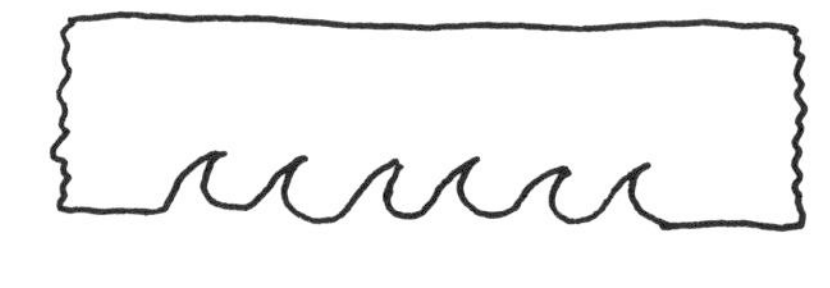

Some shapes to try

Part Two

Fact File

Unit 14

The Nature of Nails

At the end of each finger and toe there is an area of slightly curved, hardened protein called nail. Technically, nail is known as *onyx*. The protein of nail is composed of the elements carbon, hydrogen, oxygen, sulphur, and nitrogen. This protein is called keratin. It is also found in skin and hair. It is the keratin content that makes healthy nails so resiliant and pliable. These areas of keratin compare to the hooves and claws of animals.

As the cells of the new nail are formed they gradually change into hardened tissues. Normal nails fully cover the ends of the fingers and toes. Nail is colourless, has a smooth surface and is naturally flexible.

Nail forms from the clear layer of the skin of which it is an extension. The new cellular growth for nail is generated at the base of the nail.

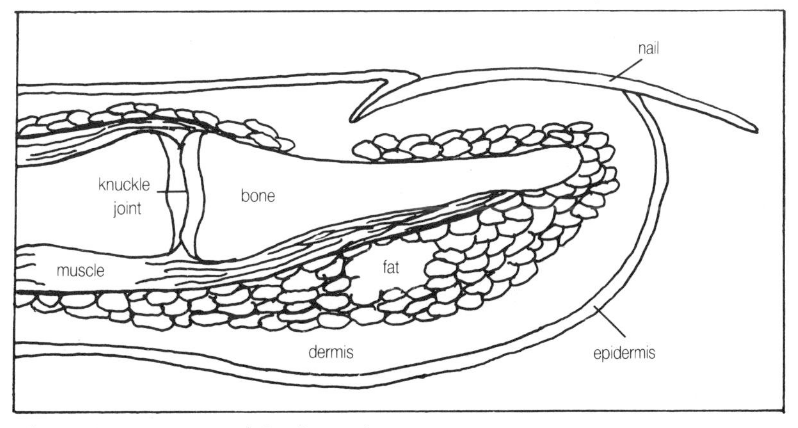

The main structures of the fingertip

NAIL FUNCTIONS

The main function of nails is to protect. They serve as a physical barrier to cushion knocks and blows, protecting the fingers and toes from damage.

Nails are not directly supplied with nerves, blood or lymph. All nail supply is from the underlying dermis. Because of this the nails can be cut without causing pain. Nails, like hair, do not bleed when cut. Yet directly beneath the nails there are many nerve endings which are sensitive to touch, pain, and so on, and there is a rich supply of blood and the necessary nutrients.

NAIL AND ADJOINING STRUCTURES

The nails are made up of a specialised form of the stratum lucidum layer of the epidermis – the translucent area of the outer layer of the skin. They protect the underlying nerves and blood vessels. They also help to support the ends of the fingers.

The nail plate is the main body of the nail, clearly visible at the ends of the fingers and toes. It is formed of translucent,

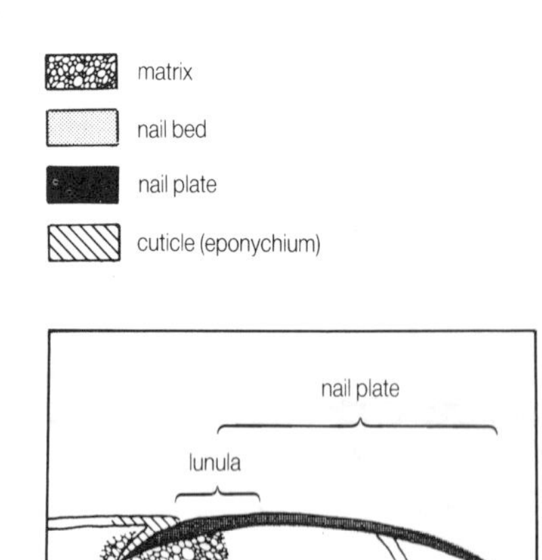

The parts of a nail, exterior

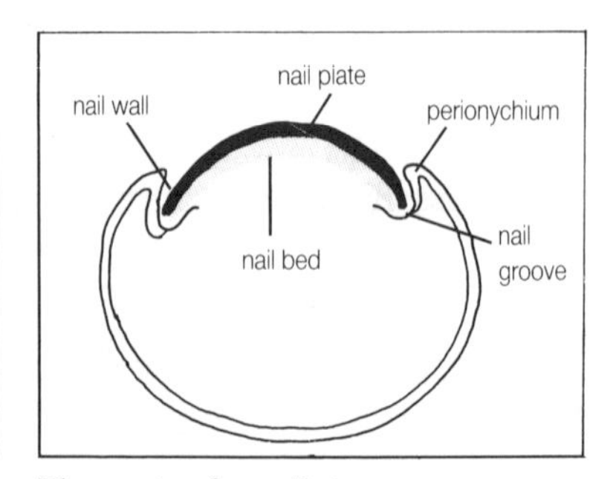

The parts of a nail, transverse section

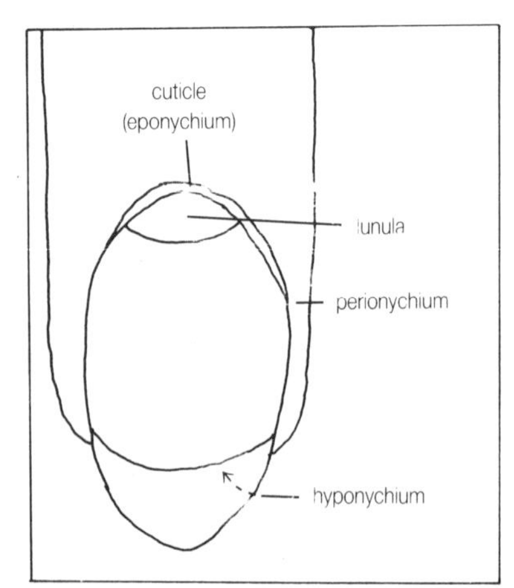

The parts of a nail, exterior

Keratinisation

This is the term given to the process of the gradual change from soft cell tissue to the hardened, firm form of nail plate tissue.

compacted, keratinised tissue. The pink colour is derived from the blood supply in the nail bed.

The nail root is at the base of the nail. It is embedded in the nail fold and surrounded by the germinating skin layer.

The free edge is the part of the nail plate that extends beyond the nail bed. Unlike the rest of the nail plate it is not in close contact with the underlying tissue, hence the lack of colour.

The matrix almost surrounds the root of the nail. This is the nail's most important feature. It is where living cells are produced that will, ultimately form the new nail plate. The matrix extends from above the last finger joint to the lunula. The cells are formed above, below and behind the root. They gradually move forwards and die, forming layers of cells held together with a small amount of moisture and fat. The stacking arrangement of the cells further strengthens the nail structure.

If the nail is damaged the nail growth is affected. If the matrix is deprived of nutrients then the nail may be malformed or have impaired growth. The matrix receives a rich supply of blood which carries oxygen – vital for cell production – and removes the waste products of cell germination.

The lunula is the lighter-coloured, semicircular area of the nail, commonly called the half moon, slightly in front of, and above, the matrix. The cells here are in the transitional stage between soft and hard – incomplete keratinisation. The lunula is always present but not always visible. The lunula may be obscured by the cuticle.

The nail bed consists of a mixed layer of skin cells. The nail plate is in direct contact with the nail bed. As the nail grows forwards, the nail bed grows with it. The nail plate does not lie in contact with the nail bed in the region of the lunula. The nail bed is richly supplied with blood vessels and nerves.

The nail walls surround three sides of the nail and are firmly attached to the sides of the nail plate.

The nail grooves are formed where the side edges of the nail plate fit the bottom of the nail walls. As the nail grows, along with the nail bed, the nails pass along the nail grooves. These guide the nails and help them grow straight.

The cuticle is a fold of skin that grows at the base of the nail, over the lunula, at the sides of the nail plate, and under the free edge. There are different names given to the different areas of cuticle.

- The **eponychium** is the dead cuticle adhering to the nail base, near the lunula.
- The **perionychium** is the cuticle that outlines the nail plate.
- The **hyponychium** is the cuticle skin found under the free edge of the nail.

The function of cuticle is to protect the areas in which it is situated.

Nail Growth

Nail growth occurs as the cells form in the matrix, change into flattened, compacted layers between the matrix and the lunula, and move between the nail walls and grooves along with the nail bed. Growth is forwards, extending beyond the fingers and toes. Nails grow constantly. New cell tissue is being generated all the time.

The nail grows approximately 3 mm ($\frac{1}{8}$th inch) per month. Nails appear to grow faster during the summer and in warm climates. Toenails grow slower than fingernails. This is possibly due to the amount of stimulation that the fingernails receive. Toenails are thicker than fingernails but all nails thicken with age.

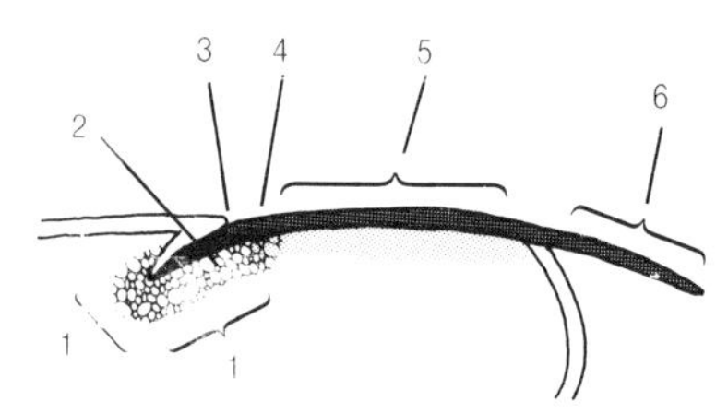

Nail growth—new cells constantly forming in the matrix keep pushing the nail plate forwards. The whole process from matrix to free edge takes about six months—about 3 mm per month

1. The matrix makes new nail cells.
2. Still in the matrix, they harden, compact, keratinise.
3. Still not fully hardened, the nail emerges from under the cuticle.
4. The lunula marks the point where the nail is fully hardened.
5. The pink* part of the nail plate is in contact with the nail bed.
6. The free edge of the nail – now you can cut it without pain.

* The colour you see through the semi-transparent nail plate is the blood-rich nail bed, which grows as the nail does.

SKIN

The skin forms a tough, flexible, outer covering for the body, that repairs itself when cut or torn. It is the largest organ of the body. Skin has the following functions:

- It produces a pigment to form a barrier against the sun's harmful rays.
- It forms a barrier against bacteria to help prevent infection.
- Extremes of heat and cold are regulated by its appendages – hair, hair muscles, sweat glands – and it helps maintain the normal body temperature of 37°C (98.6°F).
- The sweat and oil glands in the skin carry out excretion and secretion.
- Sense organs in the skin help the body respond to stimuli

such as heat, cold, pain and touch. The fingers and fingernails are well endowed with sense organs and nerve endings, enabling the effective operation of the sense of touch.

Skin Composition

The skin is composed of three main sections – epidermis, dermis and subcutaneous tissue.

Epidermis

The outermost covering of the body is the epidermis. It consists of five layers.

The outer surface is the **horny** or **cornified layer**. Beneath this is the **clear layer** through which can be seen the pink colour of the underlying blood vessels.

The next layer is the **granular layer** which is where the keratin begins to harden. This hardening process is called **keratinisation**.

Below this is the **mixed layer** of cells, containing the skin pigment **melanin**.

The lowest layer of epidermis is the **germinating layer** from which all growth stems.

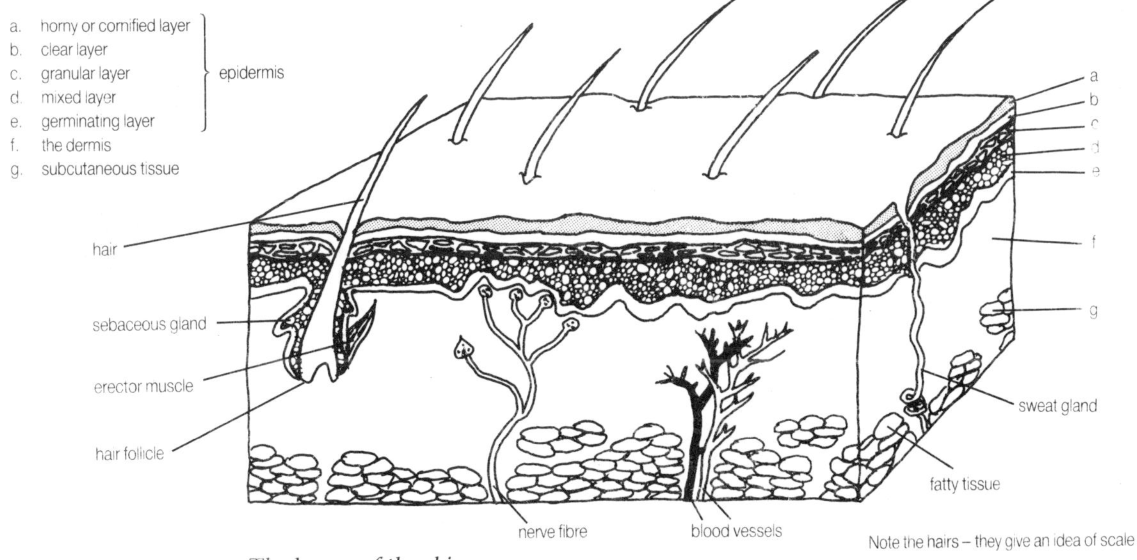

The layers of the skin

Dermis

The dermis lies beneath the epidermis. It forms the largest part of the skin. It is composed of tough, fibrous tissue. Contained in the dermis are the blood, nerve, and lymph supplies that nourish the cells, enabling them to regenerate actively and form tissues that compose the nail.

Subcutaneous Tissue

This section of the skin lies under the dermis. It is the lowest part of the skin. It contains deposits of fat, serving as a storehouse for the body. It gives shape to the body contours, fingers and toes.

MUSCLES OF THE HANDS AND ARMS

Lying in the skin, attached to the bones by tendons, are the muscles. These enable the arms and hands to move in various ways. There are muscles between the fingers but the main muscles are situated in the upper and forearms.

These consist of the **flexors** and **extensors** – responsible for lifting and straightening the arms and hands – the **supinators** and **pronators** – responsible for turning the hands palm uppermost or palms downwards – and the **abductors** and **adductors** – responsible for separating parts of the body outwards from each other or drawing parts of the body towards each other (such as separating the fingers and drawing them together).

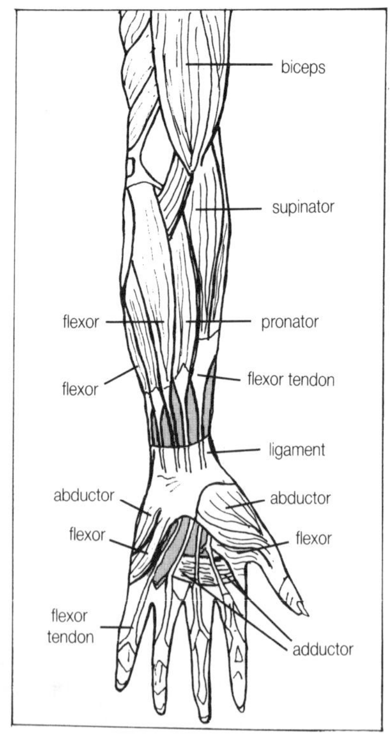

Muscles of the hand and arm — palm side

BLOOD SUPPLY

To the Arms, Hands and Fingers

The blood supply to the arms, hands and fingers is via the **radial** and **ulnar arteries**. Small capillaries convey the blood

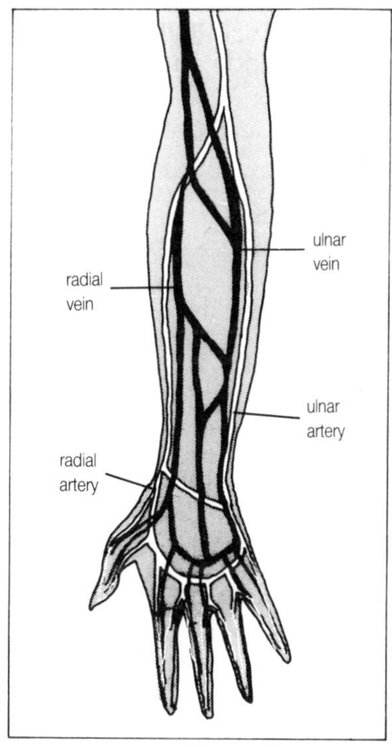

Blood supply to the hand and arm

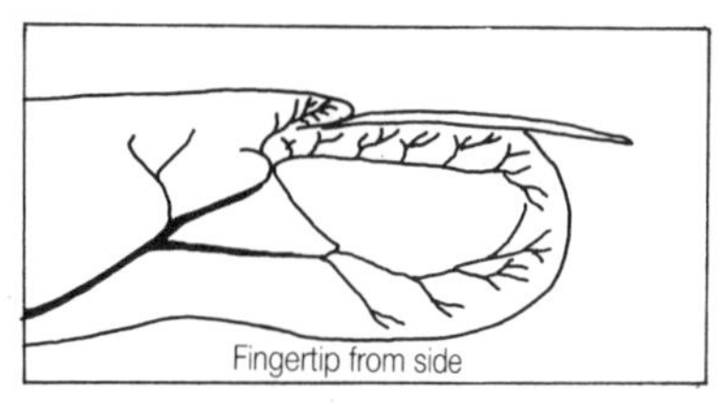

Section through fingertip shows the richness of the blood supply to the nail bed and matrix

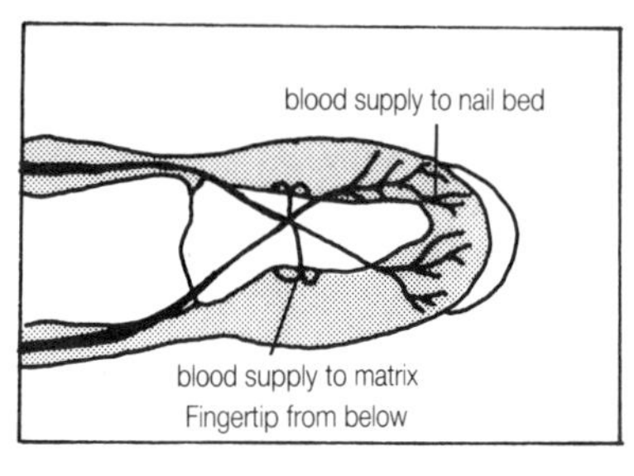

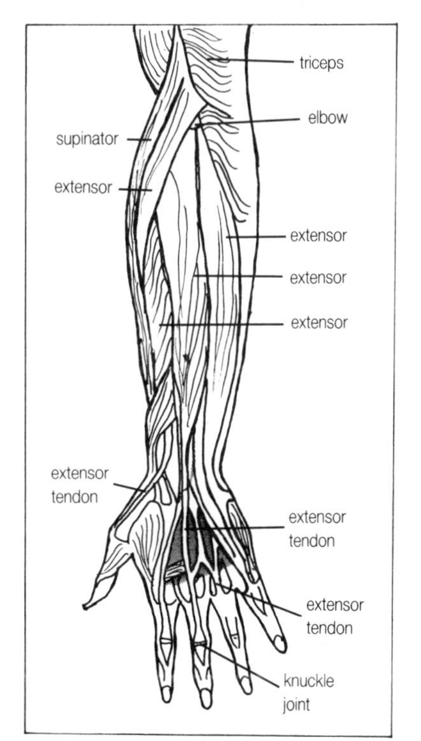

Muscles of the hand and arm—back view

through to the active growth areas, for example where nail cells are produced.

From the Fingers, Hands and Arms

Blood is returned from the fingers, hands and arms via the **radial** and **ulnar veins**. It is then conveyed through the blood system where it is oxygenated at the lungs.

NERVE SUPPLY

To the Arms, Hands and Fingers

The **ulnar nerve** and its branches, and the **median nerve** with its subdivisions, make up the main nerve supply to the arms, hands and fingers. The nerve supply enables the muscles and body to move and function. It is also responsible for sensations of touch and pain etc. to the hands and fingers.

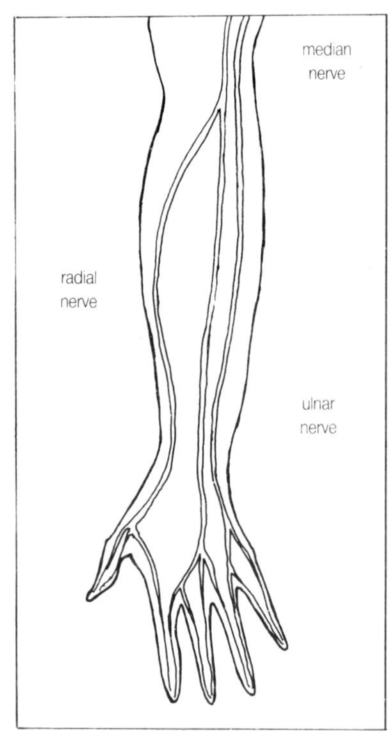

Nerve supply to hand and arm

BONE STRUCTURE

The bones of the arms, hands and fingers are as follows:

The upper arm	shoulder blades (scapula), humerus
The lower arm	ulna and radius
The wrist	eight carpal bones
The hand	five metacarpal bones
The fingers	14 phalanges on each hand (two for the thumb, three for each finger)

The bones form a framework on which other parts of the body are based. The muscles are attached to the bones by

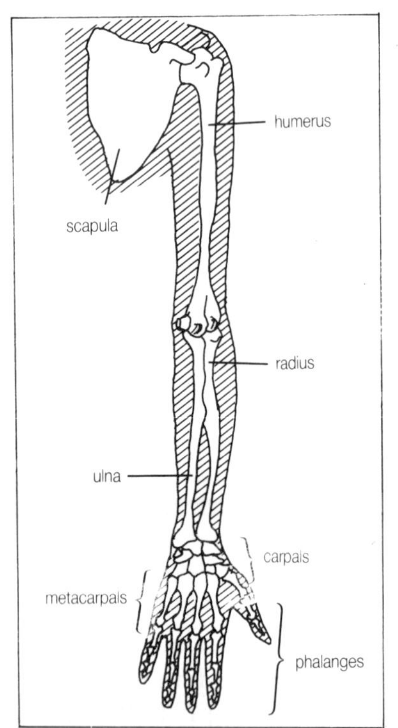

Bones of the hand and arm

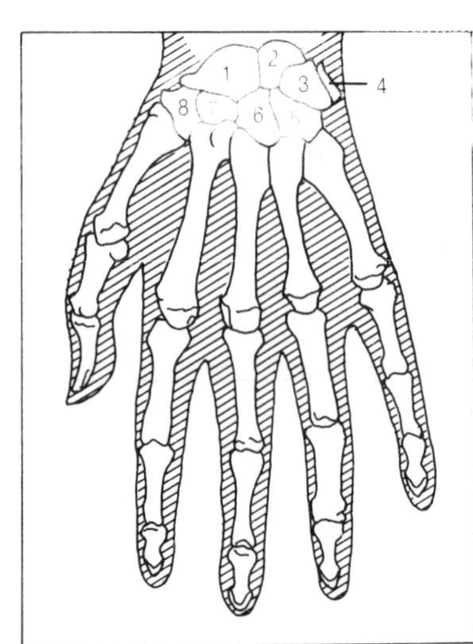
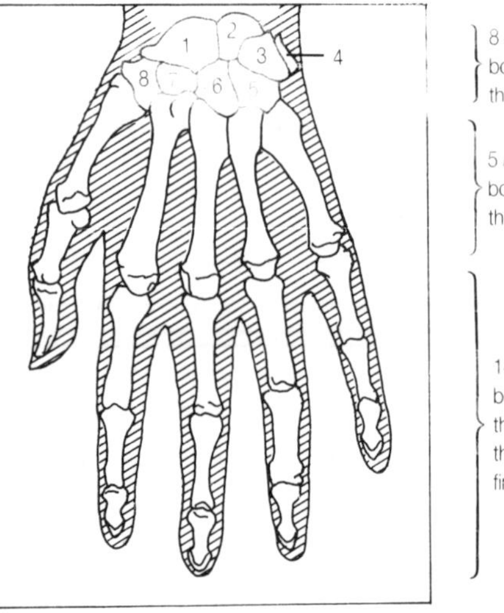

Bones of the hand

tendons which enable the various movements to take place, including the different movements of the hands and fingers.

Joints

Joints are formed where the bones meet. They are connected by strong, fibrous bands called **ligaments**. There are four kinds of moveable joint:

- **Gliding** joints – wrist, ankle
- **Ball and socket** joints – shoulder, hip
- **Hinge** joints – finger, wrist, elbow, toe, ankle, knee
- **Pivot** joints – lower arm (radius/ulna).

Moveable joints, where bones meet and move against each other, have a covering called the synovial membrane from which exudes the synovial fluid that lubricates the joints. The synovial membrane forms a sac or cavity in which the fluid is held. The knees and elbows are examples of moveable, or synovial, joints.

UNIT 15

Disorders and Diseases of Nails

Disorders of the nail are conditions caused by nail growth going wrong. Some disorders are caused by illness, others are due to physical and chemical damage, general neglect of the hands and nails and, in some cases, poor manicuring techniques.

> **Note**
>
> In most cases the appearance of the nails can be improved with professional manicure treatments and the condition controlled or stopped from recurring with appropriate home care. **Nail disorders do not contra-indicate manicure treatments but diseases do**.

Diseases of the nail are direct results of bacteria, fungi, parasites or viruses attacking the nail itself, the surrounding tissue, or both. Because of the time the nail takes to grow, the effects of the disease may well continue after the disease has been treated and cleared. The manicurist must be able to distinguish between diseases and disorders so that the correct treatment and advice can be given.

> **Note**
>
> There are risks involved with treating nail diseases. Manicure tools and linen can become contaminated ands the condition can be spread. **Nail diseases contra-indicate manicure treatment**. The client should be advised to consult the doctor for medical help.

Diagnosis is the recognition of a disease from its symptoms. Diagnostic aspects refer to signs that indicate irregularity in the body. The nails are said to reflect the state of the body. If there has been trauma, illness, or disease, then the body changes may be indicated by the condition of the nails.

If the matrix is physically damaged by injury then a distorted nail might result. The colour may range from black to blue. Interruption of the nutritional supply to the matrix can affect the nail shape, causing the formation of ridges or furrows in the nail plate.

Signs and Symptoms

The following are some of the signs and symptoms that may indicate some other condition or state of the body.

Poor Circulation

The circulation can be checked by applying pressure to and rubbing the ends of the fingers. This causes blood to leave the fingertips giving a whiter look. The quality of the circulation can be judged by how quickly the blood returns the fingertips to their normal colour. If little redness is produced and the skin remains pale, then poor circulation is indicated.

Remedies for poor circulation include the following:

- Apply or increase massage
- Buff the nails to stimulate the blood supply.

Poor circulation may indicate rheumatism or arthritis, particularly if swollen joints are interfering with the blood supply.

Thickened Nails

Thickened nails may be indicative of rheumatism, arthritis or other diseases.

Ridged Nails

Ridges in the nails may indicate illness, poor nutrition, nail damage or psoriasis. This symptom often accompanies alopecia areata, a condition producing areas of baldness on the scalp. Horizontal ridges may be due to matrix damage and will probably grow out. Vertical ridges may be due to nutritional disturbances and can be improved by buffing.

Nail Layering

The layers of the nail – called **lamellae** – often split and separate. The main cause is dryness. Layering can be the result of neglect, or incorrect nail filing rather than lack of calcium commonly thought to be the cause.

Pitting

Pitting of the nails, like ridging, can be due to psoriasis.

Matrix Damage

Damage to the matrix can be caused by incorrect manicuring, or by physical damage, such as knocks and blows, resulting in mis-shaped, ill formed nails. Nail biting is another cause.

White Spots

White spots (leuconychia) are thought to be due to air bubbles under the nail plate. These may arise as a result of matrix damage. They grow out with the nail, disappearing as they reach the free edge.

Blue Nails

This may be due to nail-bed damage. It is recognised by blood or bruising on the underside of the nail plate. Poor circulation and heart disease are other causes.

Yellow Nails

Yellow nails may be due to stains, dyes or psoriasis. Yellow patches may also reveal fungal invasion.

White Nails

Poor circulation and nail bed damage can both produce white nails. The nail may become detached.

Other Colours

Black, brown or green nails may be due to bacterial or fungal infection.

Common Disorders

Leuconychia

This is a term given to white or colourless nails or nails with white spots, streaks or bands. Leukopathia ungium is an alternative name for white or colourless nails. White colourless nails may also be a feature of albinism – a deficiency of pigment in the skin.

Cause

Leuconychia is a disorder of the nails, rather than a disease. It is due to changes in the nail plate or bed. It may be the

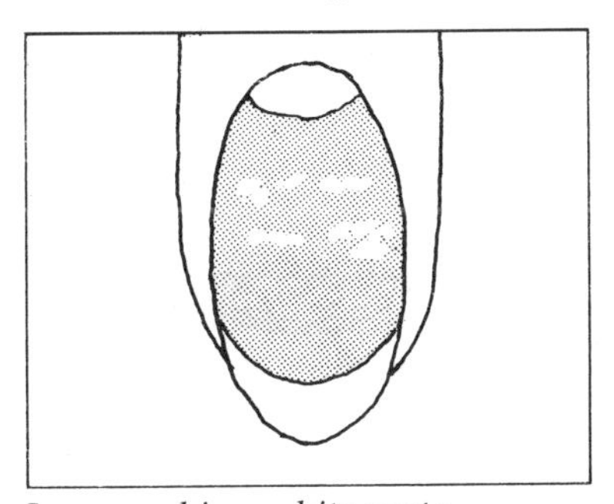

Leuconychia—white spots

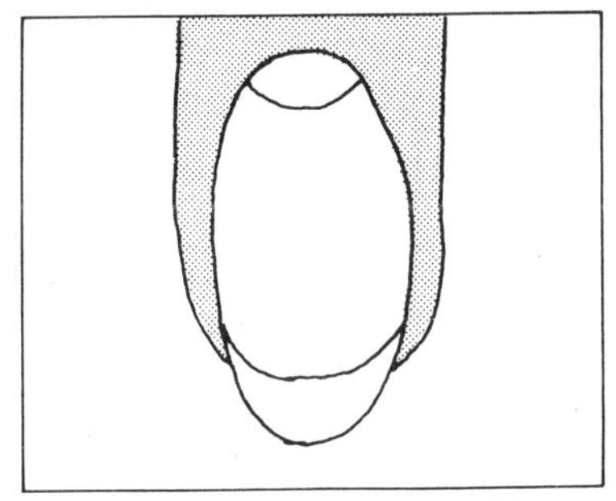

Leuconychia—completely white nail due to illness

result of injury to the matrix, or of the effects of disease. Some consider it to be due to lack of calcium or to the presence of air bubbles under the nail plate. Poor manicure techniques, for example too much pressure on the matrix, are often responsible.

Symptoms

The nails become spotted, streaked, lined or completely white. There may be evidence of ridging. Temporary or permanent whitening often occurs after extensive nail injury. Temporary white spots may be indicative of other diseases of the body.

Treatment

No treatment of the nail is needed in leuconychia. The white spots usually disappear as the nail grows.

Onychophagy

This is the technical name given to nail biting.

Cause

Nail biting is thought to be due to insecurity. Commonly the nail is chewed below the tops of the fingers. Picking toenails

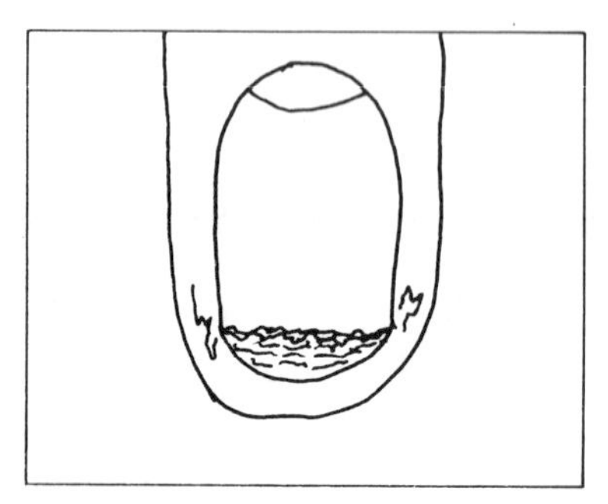

Onychophagy—bitten nails exposing nail bed

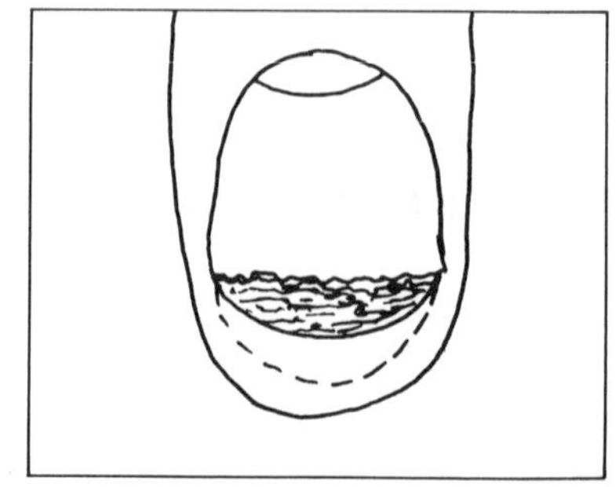

Onychophagy—flesh line recedes with repeated biting

produces similar results.

Symptoms

Nails are chewed short to a variety of degrees. The cuticle and skin at the sides of the nail are often nibbled away. Rubbing, picking, or chewing other parts of the hands may accompany nail biting. The area surrounding the nail becomes sore, painful, and inflamed.

The results of severe nail biting are often irreversible and are visible for life. The flesh line between the nail plate and the hyponychium is pushed back leaving a much lower flesh line than is normal.

Treatment

The main aspect of treatment is stopping the habit of nail biting.

Nail Ridges

Ridges in the nail – vertical and horizontal – may be due to the irregular formation of the nail, or to physical/chemical injury of the nail matrix.

Cause

Vertical ridges are common in healthy nails. They do spoil the appearance, particularly when enamelled. The cause is

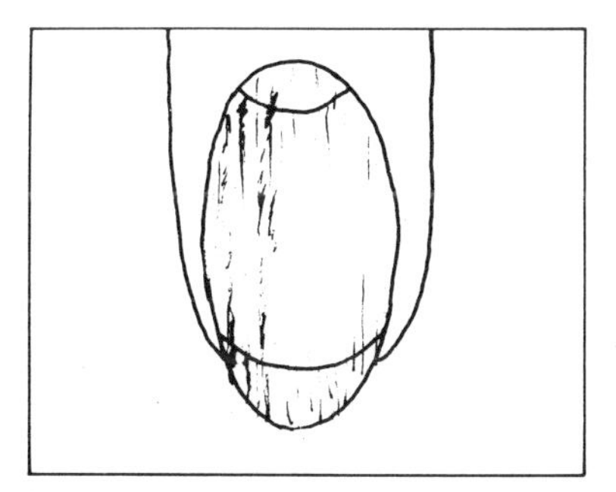

Vertical nail ridges

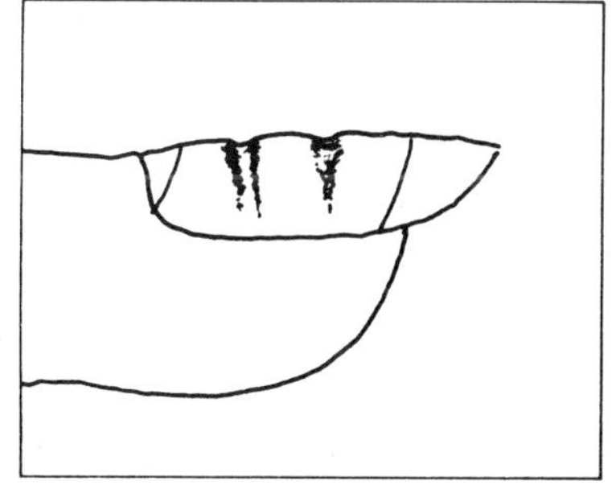

Horizontal ridges—Beau's lines

uneven development of nail tissue. This can be due to poor manicuring habits or the effects of chemicals. Buffing may minimise the appearance of shallow ridges. Deeper ones may be filled with special ridge-filler base coat before polish application.

Horizontal ridges can be an indicator of abnormal nail growth. They are often a symptom of body malfunction or disease. Deep horizontal lines/ridges are often associated with illness – for example measles or mumps – and are called Beau's lines. Ridge formation may be due to a temporary alteration of nail growth. In severe cases the nails may be shed.

Symptoms

The severity of the condition varies from shallow depressions to marked, deep ridging. There may be evidence of either physical injury or the effects of chemicals. There are several types of ridges , the type exhibited depending on the cause.

Treatment

In many cases buffing, special filling, good remedial camouflage and manicure techniques are helpful. Some ridging will grow out.

Hang Nail

There are two types of hang nail – a strip of skin that hangs loosely at the side of the nail or a small portion of the nail itself splitting away.

Cause

Hang nail may be due to dry, torn or split cuticle. Common causes are hands being immersed in water for long periods, cutting the nails too close, digging the cuticle, or improper filing. The condition may also be due to the effects of

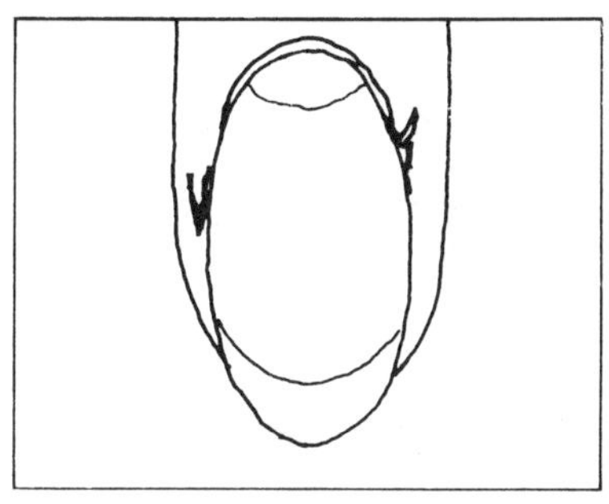

Hang nail—sliver of skin torn, sometimes at cuticle

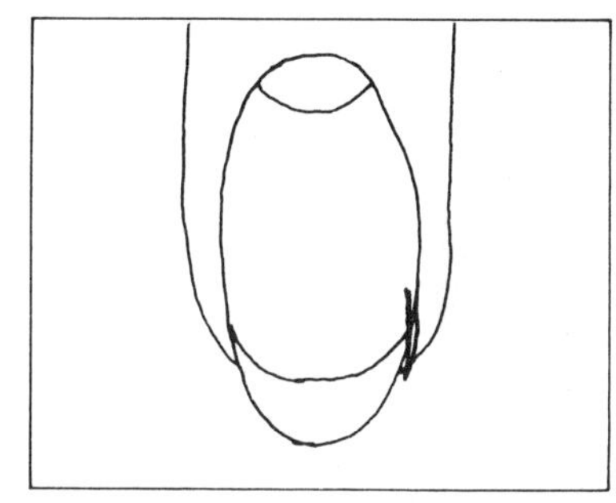

Hang nail—sliver of nail torn

detergents and other chemicals. Biting often produces hang nail.

Symptoms

The skin at the ends of the nail tends to separate, or a sliver of the nail plate tears down to expose the cutis. Inflammation and tenderness may result. If the matrix becomes affected the nail could become distorted or lost. It is important that this condition is not allowed to become severe. Early attention can stop the condition becoming further aggravated and infectious.

Treatment

Treatment may consist of oil manicures, and clipping off the cuticle with cuticle clippers to prevent pulling or chewing.

Onychogryphosis

This is the name given to an ingrowing toe or fingernail.

Cause

The condition can be caused by ill-fitting shoes, cutting or filing nails too short or too close to the skin. It may also be due to a malformation of the nail when it was beginning to grow.

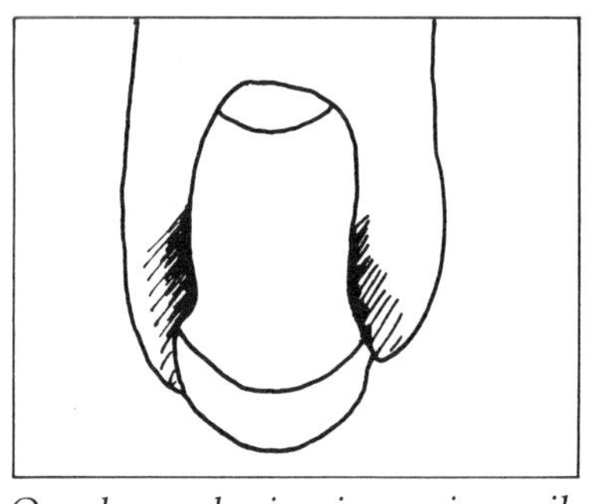
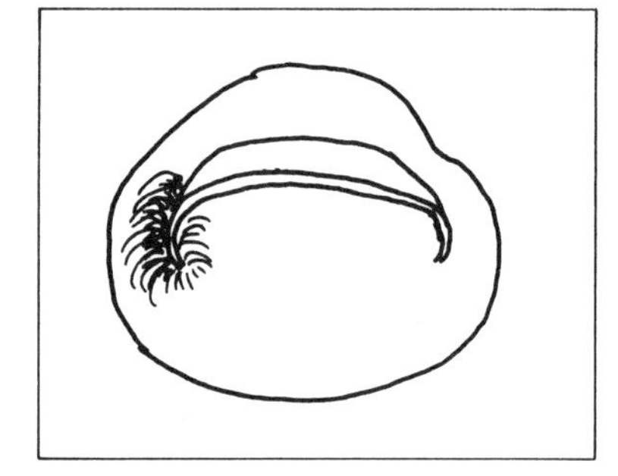

Onychogryphosis—ingrowing nail

Symptoms

The first signs are usually inflammation. This is followed by tenderness, swelling, and pain. Infection may aggravate the condition. Delaying medical attention increases the severity. The presence of onychogryphosis contra-indicates manicure.

Treatment

Medical attention is the only appropriate treatment for onychogryphosis.

Dermatitis and Eczema

Both of these terms now refer to the same condition – inflamed and irritated skin. For many years they were thought to be separate, different conditions.

Symptoms

The symptoms are inflammation, dry or moist skin, itching, swelling, and in severe states skin cracking and splitting with much pain and discomfort.

Small spots may form which break and crust to form thickened areas of skin. Water or washing appears to aggravate the condition which is best kept as dry as possible. If the condition progresses to a more severe state it is liable to become infected.

Cause

Dermatitis/eczema may be caused by factors outside the body such as chemical, physical and bacterial irritants. It may also be due to internal factors that cause allergic reaction to, for example, nickel jewellery, sticking plaster or washing-up liquids – the result is contact dermatitis.

The hands are often affected, particularly when exposed to frequent washing up or washing clothes by hand. The hands and skin become degreased and if care is not taken splits and fissures begin to appear which soon become sore.

Treatment

Dermatitis/eczema is not a condition that should be treated in the salon. Anyone affected should consult their doctor. The condition can be helped by barrier creams used before immersing hands into water, moisturing creams applied after drying the the hands, and by gently patting, rather than scrubbing, with a towel. Severe states of dermatitis/eczema contra-indicate manicure treatments or applications.

Psoriasis

Psoriasis is an abnormal thickening of the skin. The regenerative processes of skin cells appear to be increased to a rate that results in a build up of scale on the skin surface.

Cause

The cause of psoriasis is not known. There is usually a family history of the condition.

Symptoms

Psoriasis produces thick, raised, dry, silvery or white lesions. They are often circular with inflammation under and around the lesion. The lesions or spots may be formed separately or in clusters. There may be irritation. The condition varies at different times of the year – some people suffer more in the spring and summer, others in the autumn and winter. Removal of a thickened scale will reveal a tiny bleeding point.

It is not an infectious condition but a neglected state could become infected. Care and good hygiene must be used.

The nails are often affected. They may become pitted or distorted. They also may become ridged and uneven. Elbows and knees are often affected.

Treatment

Psoriasis can be difficult to treat. Medical attention is essential.

The unsightly nature of this condition causes stress to the sufferer, who is often of a nervous disposition. The manicurist can offer comfort by being tactful and understanding. Provided the condition has been attended by a doctor normal manicure procedures may be carried out. It is important that young manicurists are made aware of the nature of psoriasis and understand it is not infectious, to avoid offence and embarrassment for the client.

Pterygium

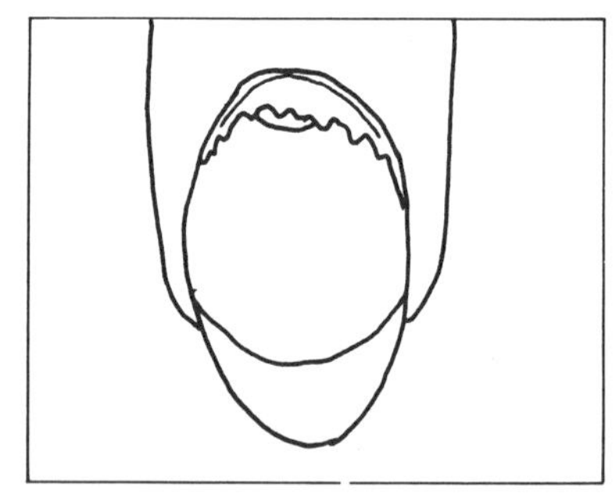

Pterygium—cuticle overgrowth

This is the name given to the condition of the nail where cuticle growth is forward and excessive.

Cause

Pterygium may be due to faulty nail care, or lack of nail care.

Symptoms

The cuticle skin at the nail base grows forward over the nail base and adheres to the nail plate.

Treatment

The cuticle knife or nail nippers can be used carefully to remove the excessive growth. Specialist manicure care is required. Oil treatment may be helpful.

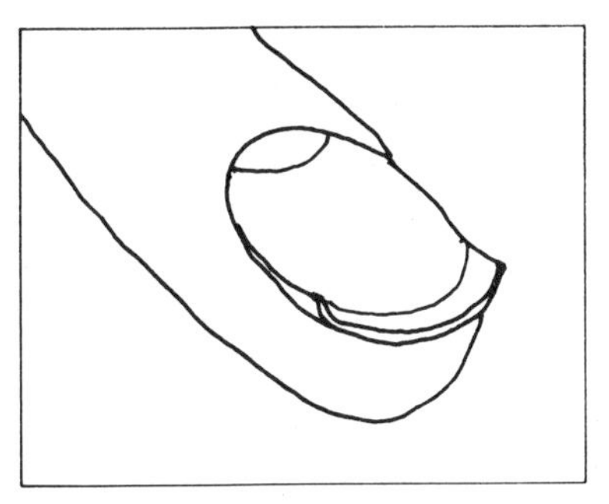

Koilonychia—spoon nail

Koilonychia

Koilonychia (spoon-shaped nails) is the name given to a concave, spoonlike nail shape.

Symptoms

The nails are thin, soft and hollowed – liquids do not run off. It is usually a secondary condition, the result of prolonged nail problems.

Cause

Koilonychia may be congenital, due to lack of iron and other minerals, or due to soft keratin. The spoon-shape results from abnormal growth at the nail matrix.

Treatment

The cause should be determined and treated. After treatment, professional manicure treatment may help.

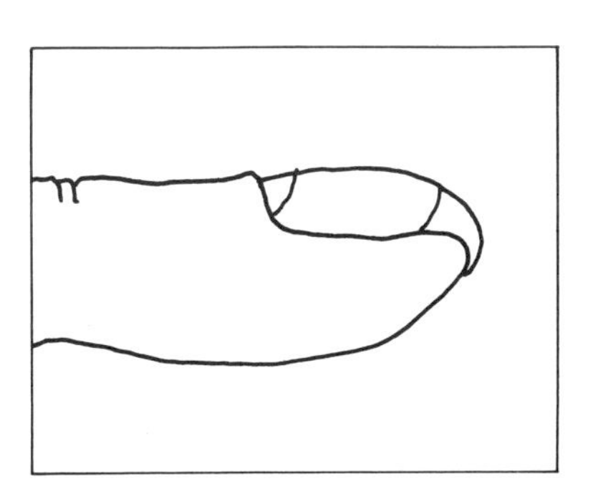

Eggshell nail

Eggshell Nails

This is the term given to thin, white nails that are more flexible than normal.

Symptoms

The nail separates from the nail bed and curves at the free edges.

Cause

The condition may be due to illness.

Treatment

After the cause of condition has been determined and treated by the doctor, professional manicure treatment may be of help.

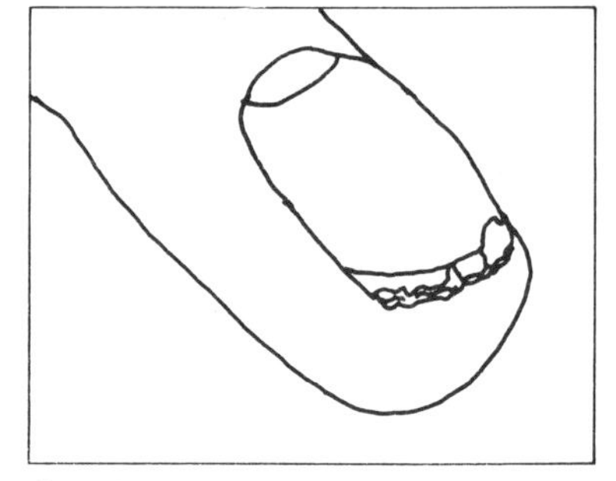
Onychorrhexis—split nails

Onychorrhexis

Onychorrhexis (fragilitis unguium) is the term given to dry, brittle nails.

Symptoms

The nail loses its moisture, becomes dry and the free edge splits. The nails may easily peel into layers. It becomes difficult to grow nails beyond the fingertips. There may be transverse or longitudinal splitting. Inflammation, tenderness, pain, swelling and infection may be present.

Cause

Frequent use of detergents, caustic chemicals and immersion in water contributes to the nails becoming dry. Natural oil protection is lost, resulting in further loss of moisture and so further dryness. Anything that dries the nail – solvents, or frequent use of nail enamel or enamel remover – encourages dryness and splitting. Incorrect filing can also cause splitting. Anaemia may be a contributing cause.

Treatments

Careful consideration to general nail maintenance, nail protection, and avoiding frequent contact with water and solvents is probably the most helpful. Good health and nutrition alleviates or helps prevent this condition.

If the condition is serious it will require medical attention. Later manicure treatments must be professionally applied.

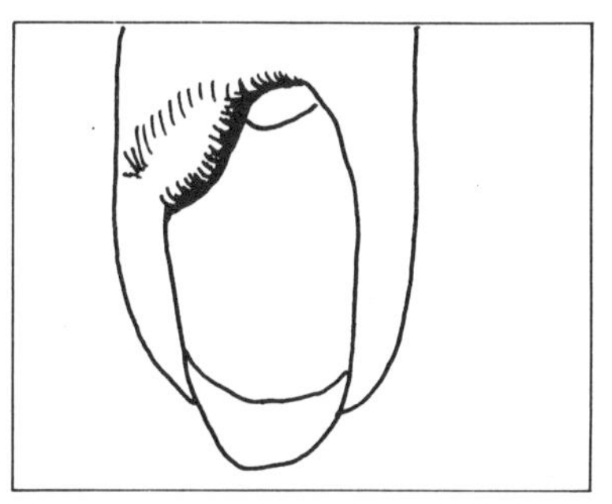

Paronychia—inflammation

Paronychia

This disease is due to inflammation of the tissue surrounding the nail, which may develop into a small abcess. It is common on the fingers.

Cause

Paronychia is caused by bacterial or viral infection. Initially, prolonged immersion of the hands in water, poor manicure techniques, picking the cuticle, and the nail wall separating from the nail, contribute to the cause.

Infection with the *Herpes simplex* virus can give rise to herpetic whitlow – an abcess that forms around the nail.

Symptoms

The tissue surrounding the nail becomes inflamed. Swelling occurs at the base and side of the nail, becoming sore, tender and painful. Pus may be formed. The nail may become damaged, distorted and loosened – perhaps finally shed. Medical attention is indicated. This disease must not be treated in the salon.

Treatment

Treatment of paronychia should be carried out only by the doctor.

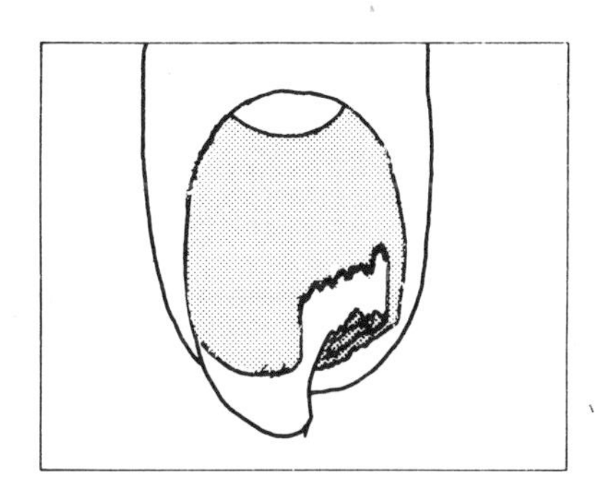

Onychomycosis—ringworm

Onychomycosis

This is the term given to fungal infections of the nail – commonly called ringworm of the nails.

Cause

Onychomycosis is due to infection of the nail, and surrounding tissue, by fungus, a vegetable parasite which attacks and feeds on keratinous tissue.

Symptoms

The free edge is usually affected first. White or yellow scaly deposits, that can be scraped off, appear on the nail plate. The nail walls or bed are then invaded. The nails may become thickened, brittle, opaque or discoloured. Onycholysis – the nail plate separating from the nail bed – is common in advanced stages. There may be accompanying dryness and skin scaling at the finger bases and on the palms.

Treatment

This condition must receive medical attention. Antibiotic cream may be prescribed.

Onychia

This term is often used to describe any disease of the nail but more specifically it refers to inflammation of the nail bed.

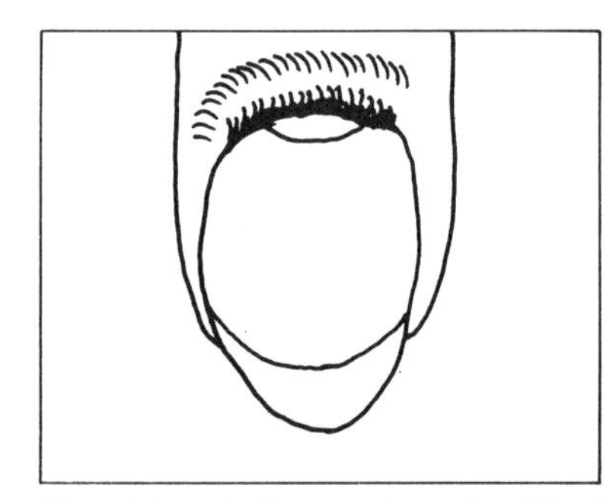

Onychia — inflammation of nail bed

Cause

The condition may be caused by wearing false nails for too long, harsh manicuring, chemical applications, a variety of infections or physical damage.

Symptoms

The nail matrix becomes red. There may be swelling, tenderness and pus formation. This could lead to the nail being shed. Unless the cause is known it should be treated as infectious.

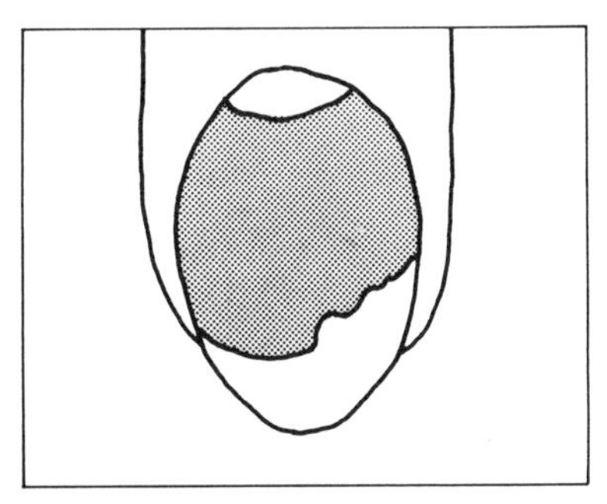

Onycholysis—lifted nail

Onycholysis

This is loosening, lifting away and separation of the nail from its bed.

Cause

It may be due to disease, physical damage, or it may occur spontaneously without any apparent cause. It can occur if sharp instruments are used under the free edge. Penetration of the flesh line allows bacteria or other infection to enter the nail bed.

Symptoms

One or more nails may become loosened. Repeated physical damage aggravates the problem. Infection commonly occurs because it is difficult to keep the area clean. It is probably the infection that makes the nail unable to re-attach itself to the nail bed. Medical treatment may be indicated. It helps to keep the nails short. Often the nail must be surgically removed.

Unit 16

Guidelines to Hygienic Manicuring

(Adapted, with kind permission, from Guidelines to Hygienic Hairdressing, *Professor N. D. Noah, Dept of Public Health and Epidemiology, King's College School of Medicine and Dentistry, London SE5 9PJ)*

TRANSMISSION OF INFECTION IN MANICURING SALONS

Manicuring is not considered a high-risk occupation for transmission of serious infection. Nevertheless a very small number of the techniques used may, in theory, pass on infection.

Manicurists and nail technicians may carry infection and transmit it to their clients. And similarly, clients may transmit infection to the manicurist. The most important risk is that of client-to-client transmission through inadequately sterilized manicure tools.

The risk of infection can be avoided altogether or kept to a minimum by good hygienic practice. Clients are becoming increasingly aware of the risks of unhygienic treatments and are demanding adequate safeguards.

INFECTIONS TRANSMISSIBLE IN MANICURING SALONS

Serious Infections

Hepatitis B and HIV (the virus that can lead to AIDS) are two very serious infections that may be transmitted by transferring small amounts of blood or other body fluids

A break in the skin, not necessarily visible, is required for transmission of HIV or hepatitis B to occur.

from an infected person or carrier (manicurist or client) to a healthy person (manicurist or client).

For safety's sake assume that every client and colleague is a potential carrier of these viruses and take the appropriate care. Transmission of infection can be avoided by the fairly simple and routine precautions outlined below.

Although the specialists say that the HIV virus does not survive for long periods on inanimate surfaces, the professional manicurist cannot rely on this. More active measures are necessary to destroy the virus. Hepatitis B virus is, in any case, much more resistant to the environment.

Less Serious Infections

Less serious infections that can be transmitted during manicure include the following:

- Spots, boils, abcesses, impetigo – streptococcal and staphylococcal infections
- Paronychia – bacterial infection
- Onycholysis – due to fungal infection
- Scabies – animal parasite infection
- Warts – viral infection
- Ringworm – vegetable parasite infection.

The methods outlined below will prevent passing on most of these infections.

AVOIDING TRANSMISSION

Metal Tools

Implements such as scissors and clippers can transmit infection if the skin of the manicurist or client is penetrated. It is not necessary for obvious bleeding to occur.

If skin is penetrated – client or manicurist – follow procedure A (treatment of wounds p. 118). Do not use the tools again until they have been treated as follows:

- Sterilized either by autoclave (procedure C p. 119) or by an approved glass-bead sterilizer (procedure D p. 120).
- If autoclave or glass bead sterilizer is not available the

tool may be boiled or steamed (procedure E p. 121).

- A permissible, though less acceptable, alternative is to soak the tool in 70 per cent alcohol, or chlorhexidine in 70 per cent alcohol, for 30 minutes – 15 minutes is the minimum (procedure B p. 118).

> **Note**
>
> Do not place contaminated tools on unprotected surfaces.

Until the tools have been treated, place them in clean paper tissue or a paper towel. Throw away the tissue or towel (procedure I p. 122) immediately after the tools have been removed for cleansing and sterilizing.

Because of the time required to treat contaminated tools, manicurists must have two or more sets available. Non-metal tools should normally be washed regularly in hot water and detergent then dried, or wiped with a fresh alcohol-impregnated wipe and allowed to dry.

Styptic

Styptic is composed of alum (aluminium potassium sulphate). It is an effective astringent and helps to stem the flow of blood from minor cuts quickly. It is made in stick, powder and aerosol forms. There is a risk of transmitting infection if styptic is used directly from a stick on to an open wound, for instance if a manicurist cuts a client's cuticle.

Never use styptic directly on broken skin. If skin is cut treat the bleeding in one of the following ways.

- Wait for bleeding to stop naturally (recommended).
- Use gauze, cottonwool or tissue on the wound, without styptic, and dispose of safely (procedure I p. 122). The wound may be rinsed first with cold water and light pressure applied.
- Rub styptic stick on to gauze, cottonwool or clean paper tissue and apply this to the bleeding wound. Dispose of safely (procedure I p. 122).
- Use styptic aerosol.

Brushes and Other Equipment

A risk arises only occasionally with these items.

Brushes

After each client, wash nail brushes in hot water and household detergent. Either dry with fresh paper towels and hand blow dryer, or use alcohol, or alcohol and chlorhexidine wipes. Do not use very stiff bristle brushes because of the danger of penetrating the skin. Brushes should not be used on broken or infected cuticles, e.g. on clients with paronychia.

Brushes in polish, cuticle remover and strengthener may be cleaned with solvents and placed in 70 per cent alcohol.

Replace worn or damaged brushes sooner rather than later.

Where possible choose products and tools that will eliminate risks of cross-infection. For instance, instead of chamois-coated, choose leather-coated buffers, which can be wiped with alcohol. Choose cuticle remover that can be applied from a tube, with a small nozzle, that never comes in direct contact with the cuticle.

Towels (and other linen)

No specific precautions are necessary except to use once only and launder after each use – preferably by boiling for five minutes or longer. This is particularly important after using on clients with cuts, bites, spots or weeping rashes on hands.

Disposable paper items – tissues, wipes, towels – are recommended for clients with skin problems (see also procedure J p. 123).

Manicure Bowls

These can be difficult to clean. They should be washed with hot water and detergent, rinsed thoroughly, and left to dry. They must be cleaned before use on another client.

Chamois Leather

This can be washed in warm soapy water but is destroyed by alcohol, boiling or autoclaving. It should be cleaned after each client.

Hands

It is important to wipe the client's hands with chlorhexidine or 70 per cent alcohol before manicure treatments to minimise the risk of cross-infection.

The manicurist's hands must be washed with soap – solid or liquid – and warm water, before and after each client. Dry hands with alcohol wipes, paper towels, or a blower.

If disinfectant hand washes are used remember they are inactivated by soap. After washing with soap, dry the hands, then apply chlorhexidine or 70% alcohol for maximum effect.

Hygiene Procedures

A Treatment of Wounds

If the client's skin has been injured, for instance by accidentally cutting the cuticle and causing it to bleed, dab the wound evenly with pre-packed spirit swab and leave to dry. Discard swab after use (see procedure I p. 122). Do not allow the wound to come in contact with other tools or equipment.

If the manicurist's skin is injured, dab with pre-packed spirit swab and cover with a waterproof dressing. Again take care not to contaminate other instruments, especially those that are difficult to sterilize, such as brushes.

If the manicurist has weeping eczema, chapped hands or other open skin problems, the hands should be protected with disposable gloves. The risk of spreading infection is not sufficient reason by itself for the manicurist to stop working or consider changing jobs. Just make sure that disposable gloves are worn and the skin condition is attended by a doctor and treated.

Infectious conditions, such as warts, must be treated by the doctor to reduce the spread of infection. Covering with plaster does not prevent cross-infection. Manicurists with warts should wear disposable gloves.

B Disinfectants

Chemical disinfectants are NOT generally recommended in manicuring salons. The exceptions are alcoholic disinfectant (see below) and bleach for blood spills (see procedure H p. 122).

Safety Tip

Alcohol is flammable. No smoking and no naked flames must be the rule when handling this and other volatile materials such as acetone.

Disinfectants readily become stale and overloaded. Some are toxic to humans and corrosive to equipment. All have to be used at the correct concentration for the correct time. Most are expensive.

Alcohol or an alcohol-based disinfectant such as chlorhexidine and alcohol can be used in certain circumstances:

- For contaminated sharp metal tools (see procedure I p. 122)
- For surfaces (see procedure H p. 122).

The recommended concentration is 70 per cent alcohol. This provides a useful bactericide though some spores are resistant. Undiluted alcohol is not effective.

When used on instruments make sure that:

- Only one or two instruments are disinfected at a time
- The alcohol covers the instruments completely
- The instruments are immersed for 30 minutes (recommended – 15 minutes minimum)
- The alcohol is discarded after ONE use.

Containers used for disinfecting with alcohol should be washed regularly with hot water and detergent, rinsed thoroughly, and left to dry.

Alcohol-impregnated wipes, available commercially, are recommended.

Discard used alcohol down the sink in running water. There are dangers when disposing of tissue or cottonwool soaked in alcohol or acetone. They must be kept in covered bins and kept well away from heat and naked flames. The fumes are toxic.

Orangewood sticks should be stored in 70 per cent alcohol between uses. Metal tools should be wiped with 70 per cent alcohol before and after use.

C Autoclaves

An autoclave is a device that sterilizes under steam pressure. If the correct temperature (higher than boiling) and pressure (higher than atmospheric) are used for the correct time (varying from 15 minutes at 121°C to three minutes at 134°C) then the instruments will be sterilized. Ordinary boiling cannot do this.

An autoclave is highly recommended, especially for

medium to large salons. It is the most efficient method of sterilizing objects. All metal objects, brushes and some plastics can be sterilized in this way. Autoclaves are easy to use, and cheap to run. And cheaper models are now available. This recommendation applies only to automatic autoclaves.

A variety of autoclaves have been approved for use by the Department of Health, for example Prestige Medical (126°C for 10 minutes).

Autoclave control indicator strips are recommended for use with autoclaves to ensure that sterilization has been achieved. These yellow strips are placed in the centre of the autoclave with the items to be sterilized. They must not be placed in the water. Sterilization is complete when the yellow changes to a purplish colour.

If at the end of the recommended sterilizing time the yellow strip has not changed colour then RESTERILIZE.

D Glass-bead Sterilizers

These are small, relatively inexpensive gadgets that sterilize by using dry heat. They can be used as an alternative to the boiling method (E) as long as the instruments are inserted for the correct length of time. After being switched on they take about 20–30 minutes to warm up, so all timings must be measured from at least 30 minutes after switching on.

Automatic timers are recommended if glass-bead sterilizers are used. The higher the temperature the shorter the time required for sterilization. But instruments are more likely to be damaged or blunted at high temperatures.

Manicurists should experiment for themselves or check with manufacturers to find out which instruments can safely be used in glass-bead sterilizers. Long pointed objects are suitable, including metal tools and scissors. Only the parts covered by the glass beads can be considered to have been sterilized. The method is not effective for awkwardly shaped items that do not make close contact with the glass beads.

The glass-bead sterilizers which have been approved for use by the Department of Health are shown in the table opposite. The times shown are the recommended minimums. It is best to sterilize each instrument separately rather than to put several in together. If more than one instrument is being sterilized at one time, the time should be increased.

Approved glass-bead sterilizers

Sterilizer	*Time* (after warm-up)
Kree	10 minutes
Nesor	10 minutes
Stericel	10 minutes
Epiltherm	3 minutes
QDendodontic	3 minutes

Note

Boiling water and steam do not sterilize – they only disinfect – because they do not reach high enough temperatures.

E Boilers/Steamers

Manicurists who use the boiler or steamer method for metal tools should use a device specially designed for instruments. These are available in most medical equipment shops.

Boiling and steaming do not sterilize because the temperatures are not high enough – they disinfect. But these methods are useful **when used properly** for cutting the risk of spreading infection.

Like the glass-bead sterilizer, boiling water or steam may be used as a temporary measure until the manicurist acquires a suitable autoclave.

The boiler/steamer should be heated by electricity and be fitted with a lid, a perforated removable shelf for raising and lowering instruments, and an automatic timer.

Instruments should be boiled or steamed for 10 minutes, then removed using clean forceps. Timing should begin when the water starts to boil – AFTER the instruments have been immersed in it.

Clean distilled or deionised water should be used and replaced completely once a day. Steamers have an automatic cutout at 10 minutes if the correct amount of water has been used. Instructions for using the steamer must be closely followed.

F Ultraviolet Light 'Sterilizers'

These are not sterilizers. Ultraviolet light has disinfectant properties only and cannot be recommended in place of any of the disinfecting and sterilizing procedures described above. Ultraviolet devices can be useful for storing clean, wrapped, sterilized tools.

G Chemical 'Sterilizing' Cabinets

These have all the disadvantages of chemical disinfectants and cannot be recommended. There is a danger from inhalation of the vapour produced by the chemicals (such as formaldehyde).

H Surfaces, Floors, Chairs

Keep basins clean using any proprietary cleanser. Other surfaces can be wiped with 70 per cent surgical spirit or similar alcohol-based disinfectant, three or four times a day. These surfaces should also be washed at the end of each day with a solution of household detergent in hot water. Floors and chairs should be kept clean with a regular wash down but need no special treatment.

Blood Spills

In the unlikely event of a blood spill, pour neat hydrogen peroxide (bleach) on to the blood, leave for a minute, and wash off with lots of hot water and detergent. Use disposable gloves. Bleach is not to be used on the client.

I Disposal

Razors and blades are sharp and special precautions must be taken for storage and disposal. All used blades and other sharp objects must be placed in a 'sharps' box. There are many inexpensive kinds available. A secure container with a screw top could also be used. When the box or container is three-quarters full it may need to be disposed of by special arrangement – discuss with your local Environmental Health Officer.

Non-sharp contaminated objects such as paper tissues and towels should be placed in a bin lined with a plastic bag. The plastic bag should be securely tied before disposal in the ordinary refuse collection system.

J Skin Rashes and Open Skin Lesions

Do not refuse to attend clients with open skin lesions such as spots, boils, abcesses, weeping eczema or rash, other than on their hands. Take the following precautions:

- Autoclave all metal tools after use
- Autoclave or boil other materials
- Wash and autoclave brushes (not plastic brushes)
- Use disposable paper towels and wipes.

If viral warts or bacterial paronychia are present on hands or fingers, do not manicure until they have cleared up.

Unit 17

The Business of Nails

This unit does not give a full business background. It deals with aspects of business that the manicurist must understand in order to offer a successful service.

RECEPTION

> You never get a second chance to make a first impression.

The reception area is of prime importance. The moment clients come through the door they begin making judgments about the service and its efficiency – before even meeting the receptionist.

Make sure that first impressions are favourable impressions.

The Reception Area

The reception area must be clean, attractive, well organised and welcoming. Comfortable seating should be provided for clients waiting for appointments. There should be a good selection of current magazines available. Tattered, dirty, outdated magazines will create the impression that the service is the same. Arrange a screened area where clients' coats and bags can be safely housed. Make sure this is always kept tidy. Battered wire coathangers on open view can look disastrous. The reception area is a suitable place to display products for resale (see selling p. 126).

A reception area

- **A.** cloakroom area
- **B.** comfortable seating
- **C.** up-to-date magazines
- **D.** product displays and advertising
- **E.** computer stores client details, list of suppliers, accounts, stock control
- **F.** appointment book

The Receptionist

The receptionist must be attractive and well groomed, with a pleasant manner and a welcoming smile. They must be calm and efficient. Clients and telephone calls should be dealt with with the minimum of delay.

The Reception Desk

This must be organised for efficiency. An appointment book should be kept that allows the scheduling of both expected and unexpected clients. Alternatively, a computer system can be used (see information technology p. 125).

Record cards must be filed tidily and accurately. These should contain essential information about the client.

The till should have a system that distinguishes between different operators, services, and retail sales (see costing manicure services p. 128) and a system for separate calculation of valued added tax (VAT). Salon owners should discuss this with their accountants who can give current up-to-date advice.

Name:		G.P.	
Address:		Tel.	
		Contra-Indications	
Tel:			
Date	Treatment	Remarks	Manicurist

Client record card

INFORMATION TECHNOLOGY

Information technology can be defined as the use of machines, such as computers and word processors, to record information. More and more businesses are using information technology to improve efficiency.

Once a computer has been installed, commercial software programs are available to meet all business requirements. **Before** installing such technology, discuss requirements with an independent expert, who can also arrange the training necessary to operate the system.

The type of functions that will be valuable include the following:

- **A computerised appointment system** allows appointments to be viewed on a display monitor. If attached to a printer this will give each manicurist an up-to-date print-out of appointments. The client can also be given a print-out of the time and date, doing away with need for appointment cards. Computers can also be linked to allow information stored in one to be transferred to another at a distant location.
- **Record cards** can be replaced by computer records which are easily updated and printed.
- **A till** linked to a computer is invaluable. It can analyse takings, and calculate VAT, the profitability of individual services, and the operators' commission.
- **Stock control** can be organised efficiently if the items entered through the till can then be debited from existing stock levels.
- **A word processing or desk top system** can produce very professional looking material and allows quick and easy communication with clients. Imagine a system that automatically informs the salon of the birthdays of clients and their children. Previous appointments can be checked. And it can be seen at a glance the most profitable time of day/week/year that appointments can be made.

  It may well be worth looking at ways that information technology can be used to improve business efficiency.

SELLING

It is essential to a manicure business that a full range of retail products are available for sale to clients.

A good product sells itself

There should never be the need to 'sell' products. The selling consists mainly of the introduction of the product and its presentation.

The clients will watch manicure procedures with interest. They will ask about the benefits of various treatments and products being used. It is important that the manicurist can give information on what is being used and how and when it is used.

Clients should be able to see the improvement in their hands and nails after a professional manicure. And they will want to maintain the improvement. Clients almost always ask to buy the products used on their hands. If they cannot buy them from the salon the nearest chemist's shop is likely to get their business. Loss of clients' business to chemists can be reduced if the products are directly available. In fact clients are likely to visit the salon more regularly – not just for the luxury of a professional manicure, but to keep their nails in shape, and importantly, to buy the products they need.

The retail range of products should be attractively displayed. The reception area is a suitable place but of course careful stock-control measures should be in operation to make sure everything is accounted for. It is interesting that in large London stores, the effectiveness of displays is judged by the number of things stolen from them by shoplifters.

A retail range should mirror every product used in the manicure treatment, as well as extra items suitable for home use. New products are constantly being developed to meet this rapidly expanding market. It is necessary to keep in close touch with these developments so that new products can be made readily available.

The following list gives an idea of the products that should be available for retail:

- Polish remover – acetone and non-acetone based
- Emery boards – wide selection from very rough to very fine
- Nail scissors – selection of sizes
- Buffing paste (see p. 23)
- Buffers – chamois-leather and four-sided (see p. 23)
- Cuticle massage cream (see p. 25)
- Cuticle oil (see p. 26)
- Cuticle remover (see p. 27)

- Cuticle nippers (see p. 30)
- Hand cream – selection of fragrances
- Base coat
- Ridge-filler base coat
- Base coat containing nail strengthener
- Coloured polishes – a very wide selection
- Top coat
- Quick-dry spray and quick-dry liquid

> **Note**
>
> Offer cream and pearlised enamels matching the colours used in the salon. These should be in smaller bottles than those intended for professional use to reduce the possibility of deterioration.

Recent additions to the range of suitable products include:

- Special nail polish remover systems where clients insert fingers into a sponge soaked in remover.
- Tubs of wipes impregnated with polish remover.
- Cuticle oil pens which have wedge-shaped tips to push back and lubricate the cuticles simultaneously.
- Cuticle trimmers which are used like a pen but have a very small blade to remove excess cuticle.
- Polish pens which allow easy polish application.
- Barrier cream to protect hands during messy jobs.
- A selection of temporary false nails in different sizes, shapes and colours, plus adhesive tabs and traditional glue.
- Nail strengtheners – both formaldehyde and non-formaldehyde sprays and oils.
- Polish remover pens with disposable wedge-shaped nibs to clean easily around the cuticle area for a perfect finish.
- Instant nail glue for easy repair of splits and breaks.

COSTING MANICURE SERVICES

It is essential that the manicurist – employer or employee – understands the principles of costing.

The employee should recognise that **time is money** and that not all the money going into the till is profit.

The employer should know the full cost of keeping the business running profitably.

Analysing costs should be the first priority and attended to before setting prices. Unfortunately, many salons base their prices on what other businesses are charging for a similar service. The clients view is often 'you get what you pay for'.

Manicure is a luxury service and if clients are satisfied it is unlikely that they will shop around for cheaper alternatives.

Underpricing is probably just as disastrous as overpricing. Charging too little results in loss. Charging too much could result in too few clients. Both result in not earning enough to operate effectively and profitably.

There are several factors to take into consideration before setting a price on the salon's services. These include:

- Wages
- Materials used
- Total overheads for running the business

Wages

If a manicurist is paid £150 per 40-hour week then it would seem at first glance that the wage cost for a manicure taking half an hour to complete could be calculated as follows:

Hourly rate = £150 ÷ 40
= £3.75
Cost for half hour = £1.88

But most operators are not working all the time, there are unproductive times to be paid for, an average of 30 hours per week is more realistic. And not every week in the year is worked. With holidays, sickness, days for training and so on the number of weeks worked will probably be nearer 45. If costs are worked over a year then the following gives a more accurate guide to cost.

£150 per week for 52 weeks = £7800 per year
Wage cost for each week worked = £7800 ÷ 45
= £173
Cost for each productive hour = £173 ÷ 30
= £5.76
Cost for half hour = £2.88

Materials

A list should be prepared of all the materials used, such as cottonwool, cuticle creams, solvents, enamels, emery boards, orange sticks and so on. Though only small amounts may be involved there is an overall cost to be arrived at. Once the amounts used have been assessed and the costs totalled, then a realistic amount for each manicure can be calculated. This contributes to the overall cost of manicuring.

Overheads

The overheads of a manicure business fall into two categories – fixed and variable.

Fixed

Fixed costs do not increase with the amount of business conducted. They include

- Rent and rates
- Interest on loans
- Light and heat
- Insurance.

Variable

Variable costs increase as the amount of business increases. They include

- Telephone, laundry, repairs, maintenance
- Cleaning and hygiene materials
- Training and updating skills
- Advertising, publicity and marketing
- Non-productive staff (juniors/receptionist/secretary/ cleaner, etc.)
- Wastage and breakage etc.

If the total overheads are arrived at and divided by the number of weeks and hours per week worked, a figure for overhead costs per hour will be arrived at. For example:

$$\begin{aligned} \text{Overheads total} &= £5000 \text{ per year} \\ \text{Cost per week} &= £5000 \div 52 \\ &= £96 \\ \text{Cost per productive hour} &= £96 \div 30 \\ &= £3.20 \end{aligned}$$

If the total costs for wages materials and overheads are

added together a basic cost of the manicure service can be assessed. It must be remembered that this is without the addition of profit, which must be included in the price charged to the client. It is usual to add a minimum of 20 per cent. This of course will be the minimum charge that should be made. All the factors involved have to be included when deciding on charges. The hidden costs of productive and unproductive hours and staff must be allowed for. The speed at which staff carry out services influences the total taken. This must be considered, and an agreed minimum of time allowed. Continuity of work should be made as even as possible throughout the working period. This involves careful booking of clients in a way that reduces the time between clients.

STOCK CONTROL

The amount of stock, or the number of products, held in a manicure salon can mean the difference between success and failure of the business.

There should be a balance between the amount of storage space used, the time that products are stored (their shelf life) and the rate at which different products are used.

A manicurist needs to understand the principles for keeping stock at a level so products do not run out but at the same time not keeping stock so long that they deteriorate.

Stock Rotation

A simple method of stock rotation is to ensure that for every product used in the salon there are two kept in the stockroom. When a product in the salon is nearly used up, one of the reserves can be taken from the stockroom. At this point it is essential to re-order the product to maintain stockroom levels.

It is important that stock received is date stamped. This ensures that old and new stock are easily distinguished. If new stock is placed behind the old it is less likely that the new stock will be used before the old. This should be done as soon as stock arrives. By rotating the stock deterioration will be kept to a minimum.

This simple system can be expanded to cover higher numbers of products – but the principle of dating and rotating stock remains. Money can be saved by buying some items in greater quantities. Care needs to be taken that only popular products are bought in this way or they will stay on the shelves for too long and exhaust their shelf life.

Accountants usually advise salon owners on how much to spend on buying and holding stock and on the total amount of stock that should be held.

Unit 18

Revision Unit

TESTS

This unit consists of a number of tests which should be useful for reminding you of what you have learnt throughout your course of training.

Your teacher/tutor/trainer will be assessing your work every now and then. You will want to know – before assessment periods – the answers to the questions that may be asked. Correct answers will make sure you pass the assessments and progress to the next section or phase of training.

There are three sorts of test.

Test 1 Asks the questions you might face in an oral assessment.

Test 2 Contains a number of sentences to be completed with the most suitable words.

Test 3 Gives a number of multiple choice questions. The answer required must be chosen from a, b, c, or d.

Self help

Use the tests after reading each unit then return to them occasionally for revision. By following this simple method you can save yourself a great deal of time and effort.

Don't forget that your clients will be asking you questions and they need reassuring that you understand what you are doing. The professional manicurist/nail technician is the one who can satisfactorily supply the answers required.

If you can answer more than half of the questions correctly you are making good progress. If fewer than half are correct return to the unit and read the text again. Where necessary check with your tutors.

If possible, work with a colleague and take turns asking each other questions. It helps to discuss the answers afterwards.

With the written questions, allow enough time to answer them accurately. Work in a quiet spot and do not refer to notes or books. Check yourself by completing the tests and then use your notebooks and texts to compare your answers.

Don't worry if it takes you a long time to get all the answers correct at first. You will make progress if you keep trying at a steady pace.

The answers to many of the questions are at the back of the book (p. 149). For other answers refer to the respective units.

Unit 3 – Patterns of Work

Test 1

1 What are the considerations to be taken if the client/manicurist is left/right-handed?
2 Name the condition where the cuticle is firmly attached to the nail plate.
3 What is the first step of the manicure procedure?
4 Explain the importance of communicating with the client.
5 Describe the sequence of manicuring.

Test 2

1 Both of the client's hands should be ________ before manicuring.
2 Polish should be ________ before progressing with the manicure.
3 If there are signs of disease manicure ________ ________ be carried out.
4 Both client and manicurist must be ________ during the process of manicuring.
5 Harsh application of manicure techniques must be avoided or ________ will result.

Test 3

1 Manicure should not be carried out if there are signs of:
   (a) nail biting (c) disease
   (b) neglect (d) disorder
2 Harsh or rough application of manicure techniques will result in:
   (a) improved growth (c) satisfaction
   (b) stimulation (d) damage
3 Before enamel application the nails should be:
   (a) clean (c) quick-dried
   (b) coloured (d) sprayed
4 Tools are placed in antiseptic to:
   (a) disinfect (c) kill all bacteria
   (b) inhibit bacteria (d) sterilize
5 Before manicure application the client's hands should be:
   (a) inspected (c) massaged
   (b) sterilized (d) polished

Test 4

List the following directions in the correct order for the completion of a manicure on a *right-handed client* (see p. 12 for answer).

(a) Soak the nails of the client's left hand.
(b) Apply buffing paste and buff the nails of the client's right hand.
(c) Soak the nails of the client's right hand.
(d) Shape the nails of the client's left hand.
(e) Remove the old polish from both hands.
(f) Apply cuticle remover to client's left hand.
(g) Inspect hands and nails to diagnose problems.
(h) Apply cuticle massage cream to the client's right hand.
(i) Loosen and push back cuticles of left hand using the appropriate tools.
(j) Remove excess cuticle from the client's right hand using cuticle clippers.
(k) Remove client's right hand from the water.
(l) Apply cuticle massage cream to the client's left hand.
(m) Shape the nails of the client's right hand.

(n) Apply the required number of coats of polish to both hands.
(o) Apply buffing paste and buff the nails of the left hand.
(p) Remove client's left hand from water and dry.
(q) Loosen and push back cuticles of client's right hand using the appropriate tools.
(r) Clean nails of both hands with nail polish remover.
(s) Apply top coat to both hands.
(t) Inspect completed work on both hands.
(u) Gently brush the nails of both hands.
(v) Hand massage both hands.
(w) Apply base coat to the nails of both hands.
(x) Apply a quick-drying spray to the nails of both hands.

Test 5

Number the fingers on the following diagrams to show the correct order for application of polish.

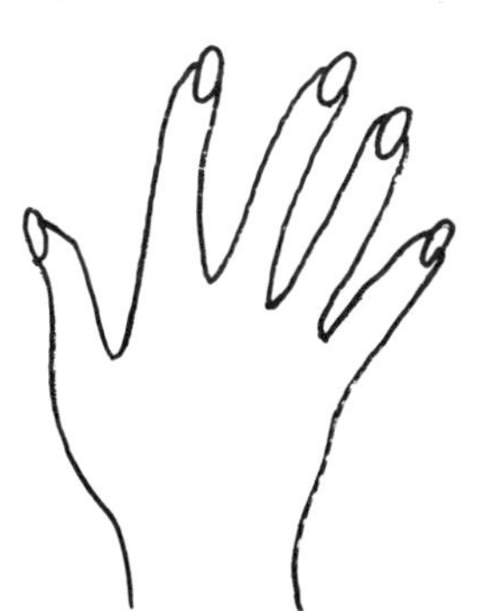

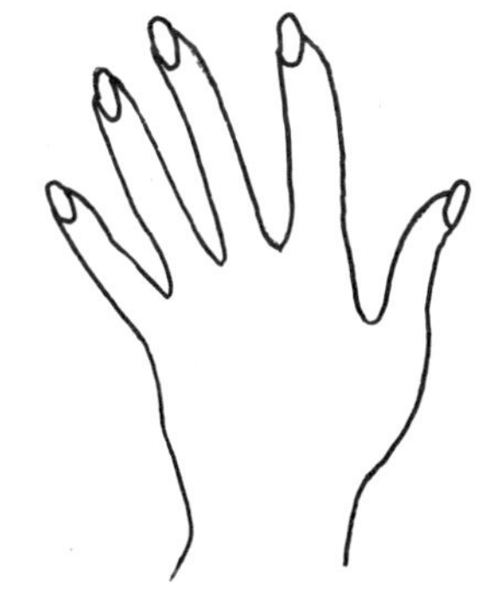

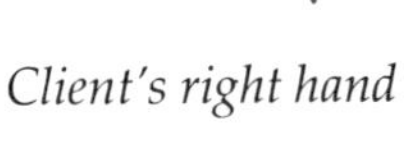

Client's right hand *Client's left hand*

Test 6

Complete the following phrases with the appropriate word/words.

Nail polishes must:

1 Be harmless to ________ and ________.
2 Be relatively easy to ________.
3 Remain stable if ________.
4 Retain ________ when dry to prevent chipping.
5 Dry ________ and ________ without creating, or trapping, air bubbles.
6 Contain no ________ vapours which could be dangerous if ________.
7 Be in a wide range of ________ to give the client plenty of choice.

Unit 4 – Shaping the Nails

Test 1

1 List seven factors to be considered when deciding the shape of the client's nails.
2 When should nail scissors be used to cut a client's nails?
3 Why is the cutting of nails not normally recommended?
4 List three rules to be followed when shaping nails with an emery board.
5 State how the nail tips should be shaped to achieve balance.

Test 2

1 The following is not a recommended shaping technique ________.
2 It is important to assess the ________ of the nail plate.
3 A slender hand is enhanced by ________ fingernails.
4 To achieve balance the ________ ________ should reflect the nail base.
5 Tools should be cleaned and ________ after use.

Test 3

1 Nails should be filed from:
- (a) centre to sides
- (b) sides to centre
- (c) backwards and forwards
- (d) one way only

2 Nails look better if they are:
- (a) square
- (b) dissimilar
- (c) similar
- (d) oval

3 Cutting nails causes:
- (a) layering
- (b) bevelling
- (c) shaping
- (d) biting

4 A main ingredient of polish remover may be:
- (a) vaseline
- (b) carbon
- (c) acetone
- (d) grease

5 The following alcohol strength is recommended for cleaning orange sticks etc:
- (a) 10 per cent
- (b) 20 per cent
- (c) 90 per cent
- (d) 70 per cent

Unit 4 – Buffing

Test 1

1 What are the beneficial effects of buffing?
2 What are the active ingredients in buffing paste?
3 List three nail conditions that will benefit from buffing treatment.
4 Draw a diagram to show the directions of the buffer on the nail plate.
5 Which type of buffer is the easiest to clean/sterilize?

Test 2

1 Buffing ________ ________ of the matrix and the nail bed.
2 When buffing do not use the buffer too ________.
3 Avoid excessive ________ when buffing.
4 Use ________ ________ when buffing to help strengthen the nails.
5 Both ________ and ________ strokes should be used when buffing.

Test 3

1 Buffing will do this to the nails:
- (a) colour
- (b) stimulate
- (c) soothe
- (d) shorten

2 Buffing paste contains:
- (a) plastic
- (b) adhesive
- (c) pumice
- (d) nylon

3 Buffers are best covered with:
- (a) leather
- (b) silk
- (c) cotton
- (d) plastic

4 A buffing paste is a mild:
- (a) abrasive
- (b) acid
- (c) adhesive
- (d) alkali

5 The use of fingers to remove buffing paste from jar causes:
- (a) psoriasis
- (b) cross-infection
- (c) disease
- (d) onychophagy

Unit 4 – Cuticle Treatment

Test 1

1 Describe how a cuticle knife is used.
2 What are cuticle nippers used for?
3 Describe how cuticle massage is given.
4 Name a strong alkali used in cuticle remover.
5 Describe how cuticle tools should be sterilized.

Test 2

1 The main purpose of cuticle massage cream is to ________ the cuticles.
2 The main reason for soaking the cuticles in warm soapy water is to ________ and ________ them.
3 A caustic alkali used in cuticle remover is ________.
4 Cuticle remover is used to remove __________ cuticle.
5 Inferior cuticle nippers are likely to ________ rather than cut the cuticle.

Test 3

1 Cuticle remover has the following action:
   (a) softening (c) polishing
   (b) reducing (d) cleaning
2 Infection is likely if an orange stick is used to:
   (a) dig (c) soften
   (b) loosen (d) free
3 This is used to clear dead cuticle:
   (a) alcohol (c) massage cream
   (b) acrylic (d) cuticle remover
4 Cuticle massage creams may contain:
   (a) pumice (c) emollient
   (b) alcohol (d) perfume
5 Cuticle remover often contains:
   (a) colour (c) alkali
   (b) polish (d) perfume

Unit 4 – Polish

Test 1

1 Why must nail enamel remover never be used to thin polish?
2 List five reasons for using a base coat.
3 Why must nail enamel be applied with the specified number of coats?
4 What gives pearlised polish its sheen?
5 What causes nail enamel to peel, chip, bubble?

Test 2

1 Old polish must be completely ________ before applying new.
2 To thin polish consistency use ________ never another polish remover.
3 A base coat helps to ________ the polish to the nail.
4 To avoid smearing/smudging polish applications should be ________.
5 Using too many polish application strokes will produce an ________ finish.

Test 3

1 Always check the following before polish application:
   (a) colour (c) age
   (b) base (d) consistency
2 The following is used in pearlised polishes to produce irridescence:
   (a) enamel (c) nitrocellulose
   (b) guanine (d) acetone
3 The following completes and protects polish application:
   (a) base coat (c) shine
   (b) colour (d) top coat
4 Polish application is made when hand massage is completed to prevent:
   (a) infection (c) drying
   (b) smudging (d) softening
5 A film of oil left on the nail may lead to polish:
   (a) keying (c) binding
   (b) peeling (d) hardening

Unit 4 – Polish Removal

Test 1

1 Why must the cottonwool soaked in polish remover be held between the index and second finger at the first joint?
2 What are the hazards of using nail polish remover?
3 Draw a diagram to show the correct direction of removal for both 1st and 2nd stages.
4 When removing polish why should rubbing see-saw fashion be avoided?
5 What is the best means of disposing of cottonwool soaked in polish remover?

Test 2

1 Fresh cottonwool may be needed for ________ nail.
2 Cottonwool should be held between the ________ and 2nd finger.
3 There are two types of remover – one for ________-based polishes and the other for nylon polishes.
4 Special care is required for ________ of cottonwool soaked with remover.
5 All used materials should be placed into ________ and placed in covered bins outside the salon.

Test 3

1 The following polish remover is more gentle on the nail plate:
(a) plastic (c) nylon-free
(b) acetone-free (d) castor oil
2 Cottonwool soaked in remover should be held between the following fingers:
(a) index and 3rd (c) index and 2nd
(b) thumb and 2nd (d) thumb and 3rd
3 The polish remover bottle should be recapped:
(a) after use (c) before use
(b) between applications (d) after applications
4 Cottonwool soaked with nail polish remover should be disposed of:
(a) afterwards (c) beforehand
(b) immediately (d) during
5 There is a remover that removes acetate-based polishes and another that removes:
(a) spirit (c) plastic
(b) alkali (d) acid

Unit 5 – Hand Massage

Test 1

1 Describe the application of massage cream.
2 Name the massage movements used.
3 How is the client's hand supported?
4 Describe finger massage.
5 Which movements complete hand massage?

Test 2

1 Five beneficial affects of hand massage are: ________, ________, ________, ________, ________.
2 ________ is a soft, smooth, stroking massage movement.
3 ________ is a deep, kneading massage movement.
4 When removing massage cream from a pot a ________ ________ should be used.
5 Complete massage by ________ ________ the client's hand.

Test 3

1 A deep kneading massage movement is called:
   (a) tapotement (c) friction
   (b) effleurage (d) petrissage
2 Remove massage cream from pots with:
   (a) spatula (c) cottonwool
   (b) fingers (d) towels
3 A smooth, soothing massage movement is:
   (a) hacking (c) effleurage
   (b) tapping (d) kneading
4 A brisk rubbing massage movement is called:
   (a) vibration (c) petrissage
   (b) effleurage (d) friction
5 Massage should always be completed with the movement called:
   (a) manipulation (c) effleurage
   (b) kneading (d) tapping

Unit 6 – Nail Repairs

Test 1

1 List the stages of nail repair.
2 Name the nail-mending materials used.
3 List the precautions to be taken with quick-setting repairers.
4 Name a common problem regarding nail repair.
5 What is the alternative to re-attaching a client's broken nail tip?

Test 2

1 Two nail mending materials are ________ and ________.
2 Film formers usually contain ________.
3 Plasticizers are added for ________.
4 Quick-setting repairers should not be used on ________ nails.
5 Examine nails before repair for ________.

Test 3

1 Nail-mending materials contain:
   (a) nitroglycerine (c) alum
   (b) nitrocellulose (d) styptic
2 A commonly used solvent is:
   (a) methylene (c) ethyl acetate
   (b) ether (d) white spirit
3 Nail polish can be applied after repair within:
   (a) 60 mins (c) 30 mins
   (b) 10 mins (d) 40 mins
4 The following may be used for nail repairs:
   (a) tissue (c) paper
   (b) lint (d) cotton
5 A silk/linen patch may be smoothed with a:
   (a) file (c) knife
   (b) emery board (d) buffer

Unit 7 – Artificial Nails

Test 1

1 State a prerequisite for using artificial nails.
2 Who are most likely to benefit from the use of artificial nails?
3 What are the precautions to be taken when using gels and glues?
4 List the artificial nails available.
5 How long should the traditional type of artificial nails be worn?

Test 2

1 Before choosing the size and shape of artificial nails it is important to ________ with the client.
2 Nail shapes can be altered by placing in ________ ________.
3 When removing traditional temporary nails ________-free remover should be used.
4 Press-on nails have the advantage of adhering ________.
5 To remove false nails soak in ________ ________ water.

Test 3

1 To remove false nails use products which are free of:
   (a) spirit (c) formaldehyde
   (b) adhesive (d) acetone
2 False nails should be worn for only:
   (a) 48 hrs (c) 12 hrs
   (b) 72 hrs (d) 24 hrs
3 Wearing temporary false nails does not cause nail:
   (a) infection (c) breakage
   (b) damage (d) growth
4 A benefit of wearing temporary artificial nails is that the nail plate is not:
   (a) bent (c) damaged
   (b) hardened (d) worn
5 When using adhesive gels and glues beware of the:
   (a) perfume (c) allergy
   (b) fumes (d) skin

Unit 8 – Semi-permanent Nails

Test 1

1 Prepare an advice sheet to be given to clients on the care and maintenance of semi-permanent false nails.
2 List the dangers of using some nail glues.
3 How should skin bonded to skin be dealt with?
4 What could be the cause of a new false nail becoming loose or detached?
5 How are semi-permanent false nails removed?

Test 2

1 Semi-permanent false nail adhesives may contain ________.
2 The adhesives used create an ________ bond between surfaces.
3 Use only ________ glues or adhesives.
4 A danger of using the wrong glue is the high ________ created.
5 Skin bonded to skin can be separated with ________ and ________.

Test 3

1 Semi-permanent false nails should remain on the natural nails no longer than:
   (a) 6 weeks (c) 5 weeks
   (b) 3 weeks (d) 4 weeks
2 For use with nail glues the following brushes are recommended:
   (a) nylon (c) sable
   (b) polythene (d) plastic
3 False nails become loose or detached mainly because of incorrect:
   (a) adhesion (c) finish
   (b) application (d) preparation
4 Nail glues may contain:
   (a) detergents (c) amyl acetate
   (b) cyanoacrylate (d) methyl methacrylic acid

5 If an artificial nail will not adhere firmly it is due to faulty:
(a) buffing (c) keying
(b) preparation (d) polishing

Unit 9 – Permanent False Nails/Extensions

Test 1

1 Outline the benefits and disadvantages of wearing permanent false nails.
2 Describe the different methods of nail adhesion.
3 Draw up a suitable advice list for client care of nail extensions.
4 What can be done when the natural nail begins to grow up?
5 How can permanent false nails be removed?

Test 2

1 ________ 'buffers' are required for shaping plastic/nylon/acrylic nails.
2 Take care when using nail clippers to avoid ________ damage.
3 Storage of plastic nails/tips should be efficient and ________.
4 Before applying a nail tip extension the natural nail surface should be lightly ________ first.
5 It is ________ for the glue to be allowed to dry.

Test 3

1 The gap between the natural and artificial nail must be filled after:
(a) 1–2 weeks (c) 3–4 weeks
(b) 2–3 weeks (d) 4–5 weeks
2 Permanent artificial nail extensions should be removed with:
(a) water (c) alcohol
(b) soap and water (d) special solvent
3 The natural nail length should not, ideally, be extended by more than:
(a) twice (c) three times
(b) half (d) three-quarters
4 Before attachment of extensions, the nail tip surface should be:
(a) smooth (c) broken
(b) small (d) roughened
5 When cutting/trimming extensions be careful of the:
(a) skin (c) cuticle
(b) eyes (d) matrix

Unit 10 – Gel and Light Systems

Test 1

1 Describe a gel ultraviolet light system for permanent false nails.
2 Temperature is important for bonding – what is the ideal?
3 Why is it important to disinfect the natural nail plate before the application of a layer of gel?
4 Discuss the differences between systems of hardening (curing).
5 What should be done if gel is accidentally 'cured' on the skin?

Test 2

1 Most primers ________ the nail plate which helps adhesion.
2 A common primer contains ________.
3 Ultraviolet light helps to cure or ________ the nails.
4 It is important to ________ the nail

before applying a layer of gel.

5 Some gels require an ________ spray rather than ultraviolet light to harden.

Test 3

1 The following chemical is found in some primers:
(a) salicyclic acid (b) methyl alcohol (c) methyl methacrylic (d) nitrocellulose

2 Ideal temperatures which assist bonding are:
(a) 11–17°C (b) 31–37°C (c) 41–47°C (d) 21–27°C

3 Before covering nails with gel clean them with non-oily polish, gauze and the following:
(a) 70 per cent alcohol (b) glycerine (c) spirit (d) disinfectant

4 The following are used in gel/light systems:
(a) IR (infra-red) (b) UVR (ultraviolet) (c) PVC (plastic) (d) TCP (antiseptic)

5 Before nail attachment, it is important that the nail plate is:
(a) massaged (b) buffered (c) trimmed (d) disinfected

Unit 11 – Maintenance and Repair of Extensions

Test 1

1 How often will a client require 'fill in' between the cuticle and the extension?

2 What is the general procedure for maintenance and repair of extensions?

3 What are glass glaze tips and what are their advantages?

4 What precautions should be taken when clipping extensions?

Test 2

1 Glass glaze tips is a method of protecting and extending nails without the use of ________.

2 Too frequent and generous use of the developing spray causes ________ and possible ________.

3 The recommended distance for the use of drying accelerator sprays is ________.

4 An advantage of using glass glaze tips is that they are ________ to use than other types of nail extension.

Test 3

1 Glass glaze tips do not require hardening with:
(a) glue (b) ultraviolet light (c) enamel (d) acetone

2 For maintenance and repair, nail extensions will need to be checked every:
(a) 2–3 days (b) 2–3 weeks (c) 4–5 days (d) 1–2 weeks

3 Product experiments are best carried out on:
(a) children (b) clients (c) men (d) models

4 The following could permanently damage nails:
(a) fungal growth (b) glass glaze (c) sharp nippers (d) polish remover

5 The following may be used to extend nails:
(a) paper (b) cottonwool (c) glass glaze (d) linen

Unit 12 – Nail Sculpture

Test 1

1 What is nail sculpture?
2 List the considerations to be taken before starting training or operating nail sculpture.
3 Describe the preparation required for nail sculpture.
4 List the precautions to be taken for nail sculpture.
5 Describe the process of filing the sculptured nail.

Test 2

1 In preparation for sculptured nails the nails should be dusted clean with an ________ ________.
2 An appropriate nail ________ has to be chosen before starting to sculpture a nail.
3 ________ filing boards are required for sculptured nails.
4 Both a liquid ________ and a powder ________ are used to form sculptured nails.
5 Adequate ________ is required when working with the sculptured nail system.

Test 3

1 The sculptured nail system involves the use of the following:
(a) acrylic resin (c) synthetic glue
(b) natural oil (d) emollient cream
2 The following is a suitable nail former:
(a) paper (c) cling film
(b) tissue (d) pre-formed aluminium
3 Because of the dangerous fumes nail sculpture should only be carried out in good:
(a) light (c) ventilation
(b) warmth (d) comfort
4 Clients should always be advised on:
(a) self help (c) nail removal
(b) after care (d) polish mixing
5 The following filing action may be used on sculptured, but not natural, nails:
(a) gentle (c) one way
(b) sawing (d) slow

Units 2–12 – General

Test 1

1 Describe an allergic reaction to nail enamel.
2 Which ingredient of nail enamel is most likely to cause an allergic reaction?
3 List three factors which should influence the choice of nail enamel colour for a client.
4 Describe the features of good quality nail enamel.
5 Explain how and when to thin down a nail enamel which has become too thick for easy use.

Test 2

1 Describe the contra-indications for nail treatments.
2 Describe the use of nail-cutting tools.
3 Name the different types of nail enamels and removers.
4 What determines the choice of the different false nails that are available?
5 List the dangers of the different nail adhesives.

Test 3

1 Outline the importance of prior consultation with the client.
2 How is a suitable enamel colour chosen?
3 List the hygienic precautions to be taken.
4 What is nail art?
5 Outline the techniques of nail shaping.

Unit 14 – The Nature of Nails

Test 1

1 What is the technical name for a nail?
2 Name three chemical elements in nails.
3 Name the protein of nails.
4 Why does the nail appear to be pink?
5 What is the main body of the nail called?
6 What is the base part of the nail called?
7 Where does nail growth take place?
8 Give two names for the nail cuticle.
9 Name five layers of the epidermis.
10 What is the semicircular area of the nail called?

Test 2

1 The ________, or half moon, is an area of incomplete keratinisation.
2 The ________ of the nail extends beyond the finger and is not attached to the skin.
3 Growth of the nails is continuous and takes place by division of cells in the ________.
4 The ________ is the portion of the epidermis under the free edge.
5 The ________ is the extension of the cuticle often overlapping the nail plate.

Test 3

1 Nail is technically known as:
(a) keratin (c) protein
(b) onyx (d) melanin
2 The actively growing area of nail is the:
(a) clear layer (c) matrix
(b) lunula (d) cuticle
3 Nail grows approximately:
(a) 6 mm per month (c) 3 mm per week
(b) 6 mm per week (d) 3 mm per month
4 Nails are not directly supplied with:
(a) keratin (c) protein
(b) blood (d) cuticle
5 Nails form from the following skin layer:
(a) cuticle (c) clear
(b) mixed (d) granular

Test 4

Name the parts of the nail indicated.

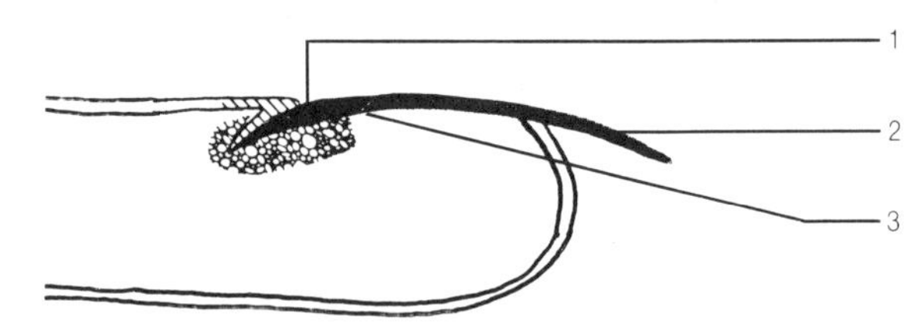

Test 5

Name the parts of the nail indicated.

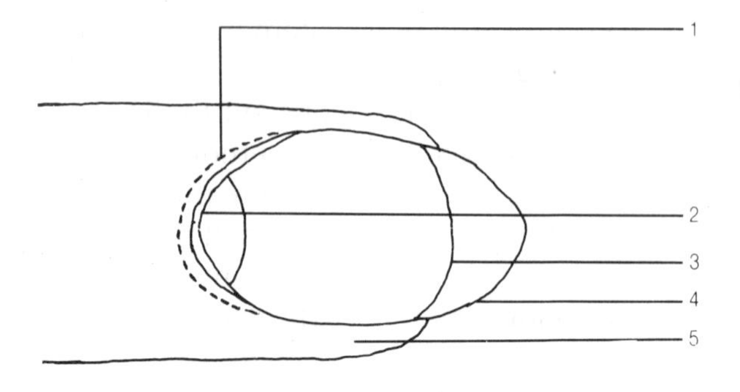

Unit 15 – Disorders and Diseases of the Nails

Test 1

1 What is the technical term for 'disease-causing'?
2 Name some disease-causing organisms.
3 Describe a non-infectious disorder of the nail.
4 Outline the symptoms and cause of a fungal nail disease.
5 What are the symptoms of paronychia?

Test 2

1 ________ is a fungal infection of the nails.
2 Destruction of all microbes is called ________.
3 Two symptoms of psoriasis are ________ ________ and a ________ ________ under the lesion.
4 ________ is a term which refers to inflammation of the nail bed.
5 Signs of infection ________ manicure treatment.

Test 3

1 The name given to a fungal infection of nails is:
   (a) onychophagy (c) leuconychia
   (b) onychomycosis (d) felon
2 Separating or loose nails are called:
   (a) onycholysis (c) paronychia
   (b) onychomycosis (d) eponychium
3 Disease-causing organisms are called:
   (a) allergens (c) non-pathogens
   (b) irritants (d) pathogens
4 The name given to white spots on the nail is:
   (a) leuconychia (c) hyperaemia
   (b) leucodermia (d) albinism
5 The following is the term given to nail biting:
   (a) onychomycosis (c) onycholysis
   (b) onychogryphosis (d) onychophagy

Test 4

1 Describe nail layering and the causes of it.
2 What is leuconychia and the possible causes?
3 Give the cause of blue nails.
4 What is onychophagy? Give causes and treatment.
5 What is indicated by nail ridging?

Test 5

1 The name given to an ingrowing fingernail is ________.
2 A strip of skin that hangs loosely at the side of the nail is called ________ ________.
3 Nail disorders are not ________.
4 Incorrect manicuring and physical damage can result in ________ damage.

Test 6

1 Nail splitting is a:
   (a) disease (c) disorder
   (b) allergen (d) pathogen
2 The technical name given to white spots in nails is:
   (a) albinism (c) onychomycosis
   (b) leuconychia (d) melanin

3 The technical name given to nail biting is:
   (a) onychophagy (c) paronychia
   (b) onyx (d) albino

UNIT 16 – HYGIENE

Test 1

1 What should be done if autoclave indicator strips do not change colour?
2 Name a disadvantage of ultraviolet cabinets.
3 What are the recommended forms of sterilization?
4 Why should styptic sticks not be used directly on the skin?
5 Why should contaminated tools not be placed on unprotected surfaces?

Test 2

1 The ________ is a recommended means of sterilizing.
2 Alcohol is ________ so no naked flames must be the rule when handling it.
3 AIDS and hepatitis are ________ infections.
4 Good hygiene reduces the risk of ________.
5 Hands should be ________ before attending a client.

Test 3

1 The following must be used at the correct temperature and pressure and for the correct time:
   (a) UVR (c) disinfectant
   (b) antiseptic (d) autoclave
2 The name given to the science of cleanliness is:
   (a) sanitisation (c) hygiene
   (b) disinfection (d) antisepsis
3 Good hygiene reduces the risk of:
   (a) cross-infection (c) disorders
   (b) disease (d) infestation
4 To minimise cross-infection clean tools may be wiped with the following:
   (a) 10 per cent alcohol (c) antiseptic
   (b) detergents (d) 70 per cent alcohol
5 Before disinfecting, all tools must be:
   (a) clean (c) brushed
   (b) sharp (d) oiled

UNIT 17 – THE BUSINESS OF NAILS

Test 1

1 What contributes to the costs of services?
2 Give examples of indirect communication.
3 Why are client records necessary?
4 How can it be ensured that products stored do not become stale?
5 What is shelf life?
6 What is meant by stock rotation?

Test 2

1 Reception desks should always be kept ________ and ________.
2 The name given to fixed and variable costs is ________.
3 Two examples of fixed costs are ________ and ________ ________.
4 An important aspect of stock control is ________ ________.
5 The term given to the use of machines

for data collection, storing, recording and retrieval is ________ ________.

Test 3

1 The receptionist should always be:
(a) happy (c) carefree
(b) efficient (d) delightful

2 A tax added to the bill for services is called:
(a) PAYE (c) capital
(b) VAT (d) sales

3 All stock removed from store must be:
(a) clean (c) stamped
(b) unopened (d) recorded

4 A manicurist paid £160 for a 40-hour week will receive the following per hour:
(a) £2 (c) £4
(b) £8 (d) £6

5 The following is a realistic number of productive working hours out of a 40 hour week:
(a) 25 (c) 20
(b) 39 (d) 30

Projects for Units 2–17

1 Prepare client advice sheets for the following:
(a) Care and maintenance of nails
(b) Dealing with false nails and frequency of salon visits
(c) Types of nail enamels available
(d) Range of materials available for home use
(e) Price lists of nail treatment – salon and home.

2 Examine the costs of the following:
(a) Manicure treatments in the salon
(b) Prices charged in different salons
(c) Goods and materials available
(d) Home treatments and materials
(e) Tools and equipment available for salon use.

3 Make lists of the following:
(a) Contra-indications to nail treatments
(b) Health and safety factors for the salon
(c) Signs and symptoms of nail diseases
(d) Prevention of cross-infection
(e) Legal requirements for salon operations

4 Outline the factors involved in each of the following:
(a) Client communication and consultation
(b) Analysis and diagnosis of clients' hands/nails
(c) Choice and reasons for selecting techniques/products
(d) Client care and maintenance
(e) Dealing with clients' complaints.

5 Examine the importance of the following:
(a) Presentation and deportment
(b) Dress and personal hygiene
(c) Maintenance of salon equipment
(d) General hygiene
(e) Speech, attitudes, disposition, body language, etc.

REVISION AIDS

The following puzzles will test how well you remember and understand some of the words you have come across in this book.

Wordsearch

The 11 answers to the questions below have been hidden in the diagram. They have been printed across (backwards or forwards), up or down, or diagonally, but always in a straight line. You can use the letters in the diagram more than once, but you don't have to use them all. You will probably find it helpful to cross the questions off the list as you find the answers.

O	R	E	S	O	L	V	E	N	T
N	V	L	A	V	F	I	L	M	E
D	A	C	K	I	L	H	I	A	C
N	P	I	G	M	E	N	T	L	A
A	C	T	L	P	X	L	E	G	R
H	O	U	T	S	O	A	A	R	P
S	D	C	X	N	R	S	R	F	U
S	R	O	S	N	E	T	X	E	S
E	A	B	D	U	C	T	O	R	S

Questions

1. Dissolves other material
2. Type of file
3. Stretch the arm outwards
4. Carries the colour in nail polish
5. Bends forearm toward shoulder
6. Protects nail root and matrix
7. Describes base coat, top coat and colourless enamel
8. Separate the fingers
9. Formed by nail polish
10. Eight wrist bones
11. Must always do this to nail-mending tissue

• Solution on p. 152.

Crossword

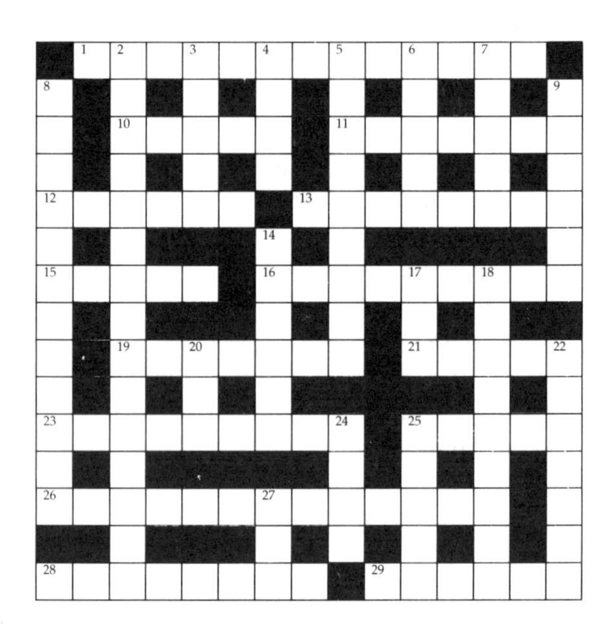

Across

1 You hit it on the head with a statue, or use it to enhance your fingertips. (4,9)
10 Eastern pork dish by the fireside, or card game. (anag.) (5)
11 Ant rail to decorate your hands. (anag.) (4,3)
12 Clint, 'e needs his nails done. (anag.) (6)
13 Ted Fince is invaded by bacteria. (anag.) (8)
15 T'alto sums it up. (anag.) (5)
16 Prime side is outside layer. (anag.) (9)
19 Polishes seal men. (anag.) (7)
21, 25 Mix to confuse (not on): something dangerous in the air. (anag.) (5,5)
23 Make Turk noise like 16. (anag.) (5,4)
25 See 21.
26 Crock lost tons of ways of managing merchandise. (anag.) (5,8)
28 Lake in L.A. has unbalanced pH. (anag.) (8)

29 Like 21, I know how to make him jump: 'Oi, son!' (6)

Down

2 P & I bemoan top-knot, and write it down here (anag.) (11,4)
3 Ken Li can point out a resemblance. (anag.) (5)
4 What a doctor does to a patient, and the light does to the gel. (4)
5 An ill song produces the growth you want. (anag.) (4,5)
6 More than once. (5)
7 Ros at Sunday lunchtime. (anag.) (5)
8 Unctuous base to deep tissue below 16. (anag.) (12)
9 Sad Dot in disagreement. (anag.) (2,4)
14 Be a floor of a ship and decorate. (6)
17 What shall we do for lunch? (3)
18 I miss exam, and he makes the most of it. (anag.) (9)
20 It's free, and you need lost of it where you find 21, 25 ac. (3)
22 Where U Chinos and make this out of a towel – under-hand! (anag.) (7)
24 Raise the tone and write it down, maybe in 2. (anag.) (4)
25 Miss Joyner with the fancy fingers. (3–2)
27 I won the right to call it mine. (anag.) (3)

- Solution on p. 152.

TEST ANSWERS

Unit 3 – Patterns of Work

Test 2

1 clean; 2 removed; 3 should not; 4 comfortable; 5 damage.

Test 3

1 (c); 2 (d); 3 (a); 4 (b); 5 (a).

Test 6

1 skin, nails; 2 apply; 3 stored; 4 flexibility; 5 quickly, evenly; 6 toxic, inhaled; 7 colours.

Unit 4 – Shaping the Nails

Test 2

1 cutting; 2 condition; 3 tapering; 4 nail tip; 5 sterilized.

Test 3

1 (b); 2 (c); 3 (a); 4 (c); 5 (d).

Unit 4 – Buffing

Test 2

1 stimulates circulation; 2 fast; 3 friction; 4 buffing paste; 5 vertical, diagonal.

Test 3

1 (b); 2 (c); 3 (a); 4 (a); 5 (b).

Unit 4 – Cuticle Treatment

Test 2

1 condition; 2 soften, loosen; 3 potassium hydroxide; 4 dead; 5 tear.

Test 3

1 (a); 2 (a); 3 (d); 4 (c); 5 (c).

Unit 4 – Polish

Test 2

1 removed; 2 polish thinners; 3 key; 4 methodical (left to right); 5 uneven.

Test 3

1 (d); 2 (b); 3 (d); 4 (b); 5 (b).

Unit 4 – Polish Removal

Test 2

1 each; 2 index; 3 acetate; 4 disposal; 5 sealed bags.

Test 3

1 (b); 2 (c); 3 (b); 4 (b); 5 (c).

Unit 5 – Hand Massage

Test 2

1 relaxes, improves blood supply, stimulates, soothes, loosens; 2 Effleurage; 3 Petrissage; 4 sterilized spatula; 5 lightly tapping.

Test 3

1 (d); 2 (a); 3 (c); 4 (d); 5 (c).

Unit 6 – Nail Repairs

Test 2

1 adhesives, solvent; 2 nitrocellulose; 3 flexibility; 4 inflamed; 5 contra-indications.

Test 3

1 (b); 2 (c); 3 (a); 4 (a); 5 (d).

Unit 7 – Artificial Nails

Test 2

1 consult; 2 hot water; 3 acetone; 4 instantly; 5 warm soapy.

Test 3

1 (d); 2 (a); 3 (a); 4 (c); 5 (b).

Unit 8 – Semi-permanent Nails

Test 2

1 cyanoacrylate; 2 instant; 3 recommended; 4 temperatures; 5 soap, water.

Test 3

1 (b); 2 (c); 3 (d); 4 (b); 5 (b).

Unit 9 – Permanent False Nails/Extensions

Test 2

1 Appropriate; 2 eye; 3 up-to-date;
4 roughened; 5 essential.

Test 3

1 (b); 2 (d); 3 (b); 4 (d); 5 (b).

Unit 10 – Gel and Light Systems

Test 2

1 etch; 2 methyl methacrylic; 3 harden;
4 disinfect; 5 accelerator.

Test 3

1 (c); 2 (d); 3 (a); 4 (b); 5 (d).

Unit 11 – Maintenance and Repair of Extensions

Test 2

1 ultraviolet light; 2 discomfort, damage;
3 12 in (300 mm); 4 easier.

Test 3

1 (b); 2 (b); 3 (d); 4 (a); 5 (c).

Unit 12 – Nail Sculpture

Test 2

1 antifungal treatment; 2 former;
3 Specialist; 4 monomer, polymer;
5 ventilation.

Test 3

1 (a); 2 (d); 3 (c); 4 (b); 5 (b).

Unit 14 – The Nature of Nails

Test 2

1 lunula; 2 free edge; 3 matrix;
4 hyponychium; 5 perionychium.

Test 3

1 (b); 2 (c); 3 (d); 4 (b); 5 (c).

Unit 15 – Disorders and Diseases of Nails

Test 2

1 Onychomycosis; 2 sterilization; 3 pitted nails, bleeding point; 4 Onychia;
5 contra-indicate.

Test 3

1 (b); 2 (a); 3 (d); 4 (a); 5 (d).

Test 5

1 onychogryphosis; 2 hang nail;
3 infectious; 4 matrix.

Test 6

1 (c); 2 (b); 3 (a).

Unit 16 – Hygiene

Test 2

1 autoclave; 2 flammable; 3 viral; 4 cross-infection; 5 washed.

Test 3

1 (d); 2 (c); 3 (a); 4 (d); 5 (a).

Unit 17 – The Business of Nails

Test 2

1 neat, tidy; 2 overheads; 3 rent, business rates; 4 stock rotation; 5 information technology.

Test 3

1 (b); 2 (b); 3 (d); 4 (c); 5 (d).

Wordsearch Solution

O	R	E	S	O	L	V	E	N	T
N	V	L	A	V	F	I	L	M	E
D	A	C	K	I	L	H	I	A	C
N	P	I	G	M	E	N	T	L	A
A	C	T	L	P	X	L	E	G	R
H	O	U	T	S	O	A	A	R	P
S	D	C	X	N	R	S	R	F	U
S	R	O	S	N	E	T	X	E	S
E	A	B	D	U	C	T	O	R	S

1 Solvent
2 Nail
3 Extensors
4 Pigment
5 Flexor
6 Cuticle
7 Clear
8 Abductors
9 Film
10 Carpus
11 Tear

Crossword Solution

Index